Rauwendaal
SPC

Chris Rauwendaal

SPC

Statistical Process Control in Injection Molding and Extrusion

2nd edtion

HANSER
Hanser Publishers, Munich • Hanser Gardner Publications, Cincinnati

The Author:
Dr. Chris Rauwendaal, Rauwendaal Extrusuion Engineering, 10556 Combie Rd # 6677,
Auburn, CA 95602, USA

Distributed in the USA and in Canada by
Hanser Gardner Publications, Inc.
6915 Valley Avenue, Cincinnati, Ohio 45244-3029, USA
Fax: (513) 527-8801
Phone: (513) 527-8977 or 1-800-950-8977
www.hansergardner.com

Distributed in all other countries by
Carl Hanser Verlag
Postfach 86 04 20, 81631 München, Germany
Fax: +49 (89) 98 48 09
www.hanser.de

The use of general descriptive names, trademarks, etc., in this publication, even if the former are not especially identified, is not to be taken as a sign that such names, as understood by the Trade Marks and Merchandise Marks Act, may accordingly be used freely by anyone.
While the advice and information in this book are believed to be true and accurate at the date of going to press, neither the authors nor the editors nor the publisher can accept any legal responsibility for any errors or omissions that may be made. The publisher makes no warranty, express or implied, with respect to the material contained herein.

Library of Congress Cataloging-in-Publication Data

Rauwendaal, Chris.
  SPC : statistical process control in injection molding and extrusion /
  Chris Rauwendaal. -- 2nd edtion.
       p. cm.
  Includes index.
  ISBN 978-1-56990-427-5
  1. Plastics--Extrusion. 2. Chemical process control--Statistical methods.
I. Title. II. Title: Statistical process control in injection molding and extrusion.
  TP1175.E9R38 2008
  668.4'12--dc22
                                                       2008009620

Bibliografische Information Der Deutschen Bibliothek
Die Deutsche Bibliothek verzeichnet diese Publikation in der Deutschen Nationalbibliografie; detaillierte bibliografische Daten sind im Internet über <http://dnb.d-nb.de> abrufbar.

ISBN 978-3-446-40785-5

All rights reserved. No part of this book may be reproduced or transmitted in any form or by any means, electronic or mechanical, including photocopying or by any information storage and retrieval system, without permission in wirting from the publisher.

© Carl Hanser Verlag, Munich 2008
Production Management: Steffen Jörg
Coverconcept: Marc Müller-Bremer, Rebranding, München, Germany
Coverdesign: MCP • Susanne Kraus GbR, Holzkirchen, Germany
Typeset: Kösel, Krugzell
Printed and bound by Kösel, Krugzell
Printed in Germany

This book is dedicated to the late Professor J. F. Ingen-Housz, a pioneer in the Dutch polymer processing field who taught for many years at the Department of Polymer Processing at Twente University, Enschede, the Netherlands. Many students benefited from the knowledge and wisdom of Professor Ingen-Housz. I feel privileged to have studied under him. In addition to being a great professor, he was a wonderful person, a true gentleman, whose warm personality and honesty were an example to all.

Preface

The basic principles of statistical process control, or SPC, were developed by Dr. Walter A. Shewhart in the 1920s. The use of SPC in American industry declined after the Second World War, while Japanese companies readily accepted and implemented SPC under the tutelage of Dr. W. Edward Deming, a former colleague of Shewhart. The success of Japanese companies in the 1970s and 1980s is at least partially a result of the widespread use of SPC. Statistical process control experienced a renaissance in American industry in the 1990s, which has contributed to a significant improvement in the country's competitive position. There is no question today that the use of SPC is an indispensable tool in world-class manufacturing operations.

This book is an updated and expanded version of the book SPC *in Extrusion* first published in 1993. The original intent was to publish a separate book on SPC in injection molding. However, with the need to update SPC *in Extrusion*, it was decided to expand the scope of the book to cover both injection molding and extrusion. One important SPC topic that was added to this book is the concept method of precontrol. This method is an alternative to the classical SPC method developed by Shewhart. Another important addition is a discussion on the Shainin methodology to design of experiments and statistical analysis.

The basic idea behind this book is to teach SPC and its application to specific processes in an integrated fashion. Many SPC training programs are taught by people who are very familiar with statistics but who know little about process technology. However, successful implementation of SPC requires an understanding of SPC as well as process know-how. This book, therefore, aims to teach not only the principles of SPC but also basic injection molding and extrusion process technology.

The first chapter deals with injection molding technology and the second chapter with extrusion technology. The third chapter discusses plastic properties that are important in molding and extrusion. In order to fully understand the process, one has to know both the machine and material characteristics. Chapter 4 is an introduction to statistical process control and Chapter 5 covers data collection, analysis, and problem solving. Chapter 6 covers measurement and Chapter 7 control charts. Chapter 8 discusses process capability and special SPC tools for injection molding and extrusion. Finally, Chapter 9 discusses design of experiments, the Shainin methodology, and precontrol.

The original book serves as the text of a companion video training program "Statistical Process Control in Extrusion," available through the Society of Plastics Engineers. The current book will form the basis of an interactive training program, "Statistical Process Control in Injection Molding," which is currently in preparation. This program will be followed by an interactive training program, "Statistical Process Control in Extrusion." These training programs will be available from Rauwendaal Extrusion Engineering as well as from the Society of Plastics Engineers.

The author would like to thank Ed Immergut, Christine Strohm, and Wolfgang Glenz for their encouragement to write this book and Martha Kürzl for managing the production. He would also like to thank the following persons for important contributions: Stan Vandercook, Peter Galuszka, Manuel Nunes, Kenny Leong, Jack Contessa, and David Hadden. The author would also like to thank Mr. Russ Nichols for drawing the author's attention to the concept of precontrol and the Shainin methodology. A book an SPC would be incomplete without these topics. The author would like to thank his children Randy, Lisette, and Yolanda for putting up with a father who spends too many hours in the study. And, last but not least, the author would like to thank his wife Sietske for her love and support in this and many other projects.

Contents

1 **Injection Molding Technology**.. 1
   1.1 The Main Components of an Injection Molding Machine................ 1
        1.1.1 The Injection Molding Cycle 2
               1.1.1.1 Characteristics of the Injection Molding Process 4
        1.1.2 The Plasticating Unit 5
               1.1.2.1 The Extruder Screw 5
               1.1.2.2 Nonreturn or Check Valves 6
               1.1.2.3 The Extruder Barrel 8
               1.1.2.4 The Nozzle of the Plasticating Unit 8
               1.1.2.5 The Feed Hopper and Feed Opening..................... 10
               1.1.2.6 The Extruder Drive 10
   1.2 The Clamping Unit 11
        1.2.1 Mechanical Clamping Systems 12
        1.2.2 Hydraulic Clamping Systems..................................... 13
        1.2.3 Hydromechanical Clamping Systems 14
   1.3 The Mold 14
        1.3.1 The Sprue 15
        1.3.2 The Runner System... 16
               1.3.2.1 Cold Runner Systems 16
               1.3.2.2 Balancing the Runner System........................... 18
               1.3.2.3 Cold Slug Traps 19
               1.3.2.4 Gates 19
               1.3.2.5 Hot Runner Systems 20
        1.3.3 Flow into the Mold 22
               1.3.3.1 Venting 24
               1.3.3.2 Weld Lines... 25
               1.3.3.3 Mold Filling Analysis 26

2 **Extrusion Technology**... 27
   2.1 Introduction 27
   2.2 The Functions of an Extruder 29
        2.2.1 Conveying 29
        2.2.2 Heating and Melting 32
        2.2.3 Mixing 34
               2.2.3.1 Distributive Mixing 36
               2.2.3.2 Dispersive Mixing 36
        2.2.4 Die Forming.. 37
               2.2.4.1 Guidelines for Shapes 39

			2.2.4.2	Guidelines for Die Design	39
		2.2.5		Devolatilization or Degassing	40
	2.3	Efficient Extrusion			41
		2.3.1		Efficient Machine Design	41
			2.3.1.1	The Extruder Screw	42
			2.3.1.2	The Extruder Barrel	43
			2.3.1.3	The Feed Hopper and Throat	44
			2.3.1.4	The Extruder Drive	46
			2.3.1.5	Instrumentation and Control	46
		2.3.2		Efficient Process Operation	47
			2.3.2.1	Feed Stock Consistency	47
			2.3.2.2	Temperatures	48
			2.3.2.3	Screen Pack	49
			2.3.2.4	Feeding	50
			2.3.2.5	Gear Pumps	51

3 Plastics and Plastics Properties Important in Injection Molding and Extrusion — 53
   3.1 Thermoplastics and Thermosets — 53
   3.2 Amorphous and Semicrystalline Plastics — 53
   3.3 Liquid Crystalline Plastics — 54
   3.4 Elastomers — 55
   3.5 Flow Behavior of Plastics — 55
      3.5.1 The Melt Index Test — 56
      3.5.2 Spiral Length Tester — 57
      3.5.3 The Effect of Shearing — 57
         3.5.3.1 Shear Thinning or Pseudoplastic Behavior — 58
         3.5.3.2 Effect of Temperature on Viscosity — 59
         3.5.3.3 Effect of Pressure on Viscosity — 59
      3.5.4 Flow Properties for Injection Molding — 59
      3.5.5 Viscous Heat Generation — 60
   3.6 Thermal Properties — 60
      3.6.1 Thermal Conductivity — 61
      3.6.2 Specific Heat and Enthalpy — 61
      3.6.3 Thermal Stability and Induction Time — 62
      3.6.4 Density — 62

4 Introduction to Statistical Process Control — 65
   4.1 Introduction — 65
   4.2 Implementing Statistical Process Control — 68
   4.3 Basic Statistical Concepts — 73
      4.3.1 Causes of Variability — 73
      4.3.2 Basic Statistical Terms — 75

	4.3.3	Mean, Median, and Mode	75
	4.3.4	Range, Variance, and Standard Deviation	76
	4.3.5	Plotting Distribution Patterns	78
	4.3.6	Characteristics of a Frequency Distribution	82
	4.3.7	Different Distribution Patterns	83
	4.3.8	The Central Limit Theorem	84

5 Data Collection, Data Analysis, and Problem Solving … 87

- 5.1 Attributes Data Versus Variables Data … 87
- 5.2 Important Aspects of Data Collection … 87
- 5.3 Diagrams for Problem Solving … 88
  - 5.3.1 Cause Listing Diagram … 88
  - 5.3.2 Variation Analysis Diagram … 89
  - 5.3.3 Process Analysis Diagram … 90
- 5.4 Pareto Diagrams … 91
- 5.5 Histograms … 93
- 5.6 Scatter Diagrams and Correlation Tables … 93
- 5.7 Recording Data … 95
  - 5.7.1 Check Sheets … 95
  - 5.7.2 Portable Data Collectors/Machine Analyzers … 97
  - 5.7.3 Fixed Station Data Acquisition Systems … 98
- 5.8 Sampling … 101

6 Measurement … 103

- 6.1 Introduction … 103
- 6.2 Basic Concepts … 104
- 6.3 Measurement Error … 106
- 6.4 Quantifying Measurement Variation … 107
  - 6.4.1 Short Method for Gage R&R … 109
  - 6.4.2 Long Method for Gage R&R … 110
  - 6.4.3 Gage Accuracy … 113
  - 6.4.4 Gage Stability … 113
  - 6.4.5 Gage Linearity … 114
    - 6.4.5.1 Guidelines … 114
- 6.5 Graphical Method for Measurement Analysis … 114
  - 6.5.1 Traditional Gage R&R Method … 115
  - 6.5.2 Graphical Approach to Gage R&R … 116
  - 6.5.3 Differences Between Operators … 118
    - 6.5.3.1 Measurement Error and Product Variation … 118
- 6.6 Measurements in Injection Molding … 120
  - 6.6.1 Important Process Parameters … 120
    - 6.6.1.1 Product Characteristics … 121

			6.6.1.2	Pressure	121
			6.6.1.3	Temperature	124
			6.6.1.4	Screw Speed	126
	6.7	Measurements in Extrusion			126
		6.7.1	Important Process Parameters		126
			6.7.1.1	Extrudate Dimensions	127
			6.7.1.2	Pressure	127
			6.7.1.3	Screw and Take-Up Speed	129
7	**Control Charts**				131
	7.1	Introduction			131
	7.2	Control Charts for Variables Data			131
		7.2.1	The $\bar{x}$ and R Chart		132
	7.3	Interpretation of $\bar{x}$ and R Charts			137
	7.4	Control Charts for Attributes Data			140
		7.4.1	Creating and Analyzing Attributes Control Charts		141
		7.4.2	The p Chart		141
		7.4.3	The np Chart		144
		7.4.4	The c Chart		146
		7.4.5	The u Chart		147
8	**Process Capability and Special SPC Techniques for Molding and Extrusion**				151
	8.1	Introduction			151
	8.2	Capability Indices			152
		8.2.1	Process Capability Shortcut with Precontrol		154
	8.3	Tests for Normality			155
		8.3.1	Graphical Methods		155
		8.3.2	Statistical Calculations		157
	8.4	Capability Measures for Non-Normal Distributions			159
	8.5	Special SPC Techniques for Injection Molding			161
		8.5.1	Family Processes		161
			8.5.1.1	Median/Individual Measurement Control Charts	162
			8.5.1.2	Group Charting	164
		8.5.2	Grading Machine Capability		165
	8.6	Example of SPC in Extrusion			167
		8.6.1	Introduction		167
		8.6.2	Specific Example Overview		168
		8.6.3	Low Cost Portable SPC Package		169
		8.6.4	The Extrusion Line		170
		8.6.5	Extrusion Line Changes, General		171
		8.6.6	Extrusion Line Changes, Details		172
		8.6.7	Conclusions		173

9	**Other Tools to Improve Process Control**	175
9.1	Introduction	175
9.2	Design of Experiments (DOE)	176
9.3	The Shainin Methodology	177
	9.3.1 Tools to Generate Clues	179
	9.3.1.1 Multi-Vari Charts	179
	9.3.1.2 Components Search	180
	9.3.1.3 Paired Comparisons	182
	9.3.1.4 Variables Search	182
	9.3.2 Factorial Designs	184
	9.3.2.1 Interaction	185
	9.3.2.2 Two-Level Factorial Design with K Factors	187
	9.3.3 The B Versus C Analysis	188
	9.3.4 Scatter Plots to Determine Appropriate Tolerances	192
9.4	Precontrol	194
	9.4.1 Comparing Precontrol to Control Charts	196

References ... 199

Appendix I, List of Polymer Acronyms ... 203

Appendix II, Nomenclature ... 213

Appendix III, SPC/DOE Software ... 215

Appendix IV, Expressions for the Normal Distribution Curve and z-Table ... 223

Appendix V, Glossary ... 227

Appendix VI, Monitoring Systems for Injection Molding ... 235

Appendix VII, Conversion Constants ... 237

Appendix VIII, List of Acronyms ... 241

Appendix IX, Proposed New Terminology ... 243

Subject Index ... 245

1 Injection Molding Technology

1.1 The Main Components of an Injection Molding Machine

The main units of a typical injection molding machine are the clamping unit, the plasticating unit, and the drive unit; they are shown in Fig. 1.1. The clamping unit holds the injection mold. It is capable of closing, clamping, and opening the mold. Its main components are the fixed and moving platens, the tie bars, and the mechanism for opening, closing, and clamping.

The injection unit or plasticating unit melts the plastic and injects it into the mold. The drive unit provides power for the plasticating unit and clamping unit.

Injection molding machines are often classified by the maximum clamp force that the machine can generate. This is the force that pushes the two mold halves together to avoid opening of the mold due to internal pressure of the plastic melt in the mold. In the US the clamp force is usually expressed in tons (US); one ton equals 2000 poundforce (lbf). In other countries the clamping force is usually expressed in units of the SI system, the international system of units. This could be kilonewtons or metric tons; 1 metric ton = 1000 kilogram-force (kgf) = 9.81 kilonewtons (kN) = 1.1 US tons. The clamping force of typical injection molding machines range from 200 to 100,000 kN (about 20 to 10,000 metric tons).

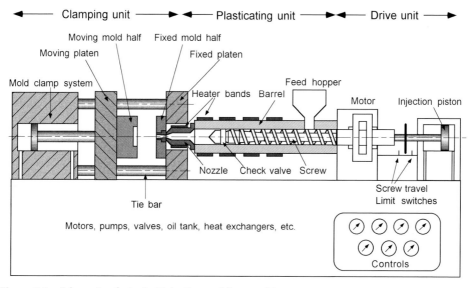

Figure 1.1 Schematic of a typical injection-molding machine

In some cases the maximum injection pressure is of interest. The maximum injection pressures range from 150 to 250 megapascals (MPa); this corresponds to about 22,000 to 36,000 psi. Another machine parameter of importance is the maximum injection capacity. In metric units this is often given in cubic centimeters (cc or cm^3), in the US it is often given in cubic inches or ounces of polystyrene; 1 cc = 0.061 inch^3 and 1 ounce = 28.3 grams. To convert from ounces of PS to another plastic, for example LDPE, the number has to be multiplied with the density ratio of LDPE to PS. The density of PS is about 1.06 gr/cc and of LDPE about 0.92 gr/cc. For an explanation of abbreviations, see the polymer acronym list in Appendix I.

Another machine parameter that is given in some cases is the diameter of the screw of the plasticating unit. The typical range of screw diameters is from 20 to 120 mm (0.8 to 4.7 inches), although larger diameters are used for some very large injection molding machines.

1.1.1 The Injection Molding Cycle

There are three main stages in the injection molding cycle: stage 1, injection, followed by stage 2, holding pressure and plasticating, and finally, stage 3, ejection of the injection molded part. When stage 3 is completed, the mold closes and the cycle starts over again.

Stage 1, Injection of the plastic melt into the mold. In stage 1, the mold is closed and the nozzle of the extruder is pushed against the sprue bushing of the mold. At this point the screw, not rotating yet, is pushed forward so that the plastic melt in front of the screw is forced into the mold; see Fig. 1.2.

Stage 2, Holding pressure and plasticating. When the mold is completely filled, the screw remains stationary for some time to keep the plastic in the mold under pressure; this is called the "hold" time. During the hold time additional melt is injected into the mold to compensate for contraction due to cooling. Later, the gate, which is the narrow entrance into the mold, freezes. At this point the mold is isolated from the injection unit; however, the melt within the mold is still at high pressure. As the melt cools and solidifies, the

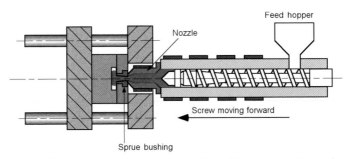

Figure 1.2 Stage 1 of the injection molding cycle: injection of the plastic melt into the mold

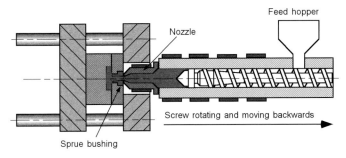

Figure 1.3 Stage 2 of the injection molding cycle: holding and screw recovery

pressure should be high enough to avoid sinkmarks, but low enough to allow easy removal of the parts.

When the gate freezes, the screw rotation is started. The period of screw rotation is called screw "recovery;" see Fig. 1.3. The rotation of the screw causes the plastic to be conveyed forward. As the plastic moves forward, heat from the barrel and shear starts to melt the plastic. At the discharge end of the screw, the plastic will be completely melted. The melt that accumulates at the end of the screw pushes the screw backward. Thus the screw is rotating and moving backward at the same time. The rate at which plastic melt accumulates in front of the screw can be controlled by the screw backpressure, that is, the hydraulic pressure exerted on the screw. This also controls the melt pressure in front of the screw.

When sufficient melt has accumulated in front of the screw, the rotation of the screw stops. During screw recovery the plastic in the mold is cooling, but typically the cooling is not finished by the end of screw recovery. As a result, the screw will remain stationary for some period until cooling is completed. This period is often referred to as "soak" time. During this time additional plastic will melt in the extruder from conductive heating. Also, the melted material will reach more thermal uniformity, although the soak time is usually too short to improve thermal homogeneity significantly.

Stage 3, Ejection. When the material in the mold has cooled sufficiently to hold its shape, the mold opens and the parts are ejected from the mold; see Fig. 1.4. When the molded part has been ejected, the mold closes and the cycle starts over again.

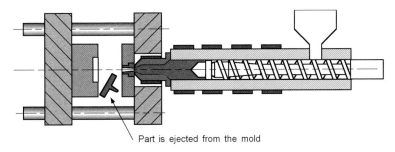

Figure 1.4 Stage 3 of the injection molding cycle: ejection of the part(s)

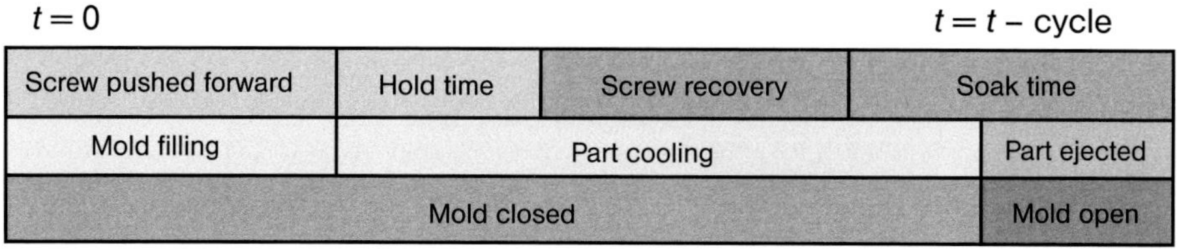

$t = 0$			$t = t - \text{cycle}$
Screw pushed forward	Hold time	Screw recovery	Soak time
Mold filling	Part cooling		Part ejected
Mold closed			Mold open

Figure 1.5 The injection molding cycle

The different stages can be graphically illustrated as shown in Fig. 1.5. The top bar shows the movement of the extruder screw, the second bar shows the action going on inside the mold, the third bar indicates at what times the mold is open and closed.

As can be seen in Fig. 1.5, the major part of the injection molding cycle is the cooling time required for the plastic in the mold to reduce to a temperature at which the part can be removed without significant distortion. The main variable that determines the cooling time is the thickness of the molded part. A simple relationship can be used to determine the approximate time required for cooling:

cooling time = $2(\text{wall thickness})^2$

In this expression the cooling time is expressed in seconds and the wall thickness in millimeters (1 mm = 0.001 m). For example, if the wall thickness is 1 mm, the cooling time will be about 2 seconds. With thin parts, it is possible to achieve cycle times of just a few seconds. An example of such a part is the compact disk or CD. If the wall thickness of the part is 5 mm, the cooling time will be about 50 seconds. In this case, the cycle time will be about one minute. It is clear, therefore, that the cycle time increases strongly when the wall thickness increases.

1.1.1.1 Characteristics of the Injection Molding Process

The injection molding process is characterized by:

1. Expensive molds: as a result, the injection molding process is suitable only for large production runs.
2. Low assembly costs: these are made possible by the ability to manufacture complicated, integrated products.
3. High pressures generated in the injection molding machine: this limits the size of the products that can be made to approximately 1 m^2.
4. The maximum ratio of flow length L to thickness H of about 300: as a result, long products require multiple gates.
5. Maximum part thickness of about 5 mm: this is necessary to keep the cooling time within reasonable limits; product stiffness can be improved by the use of ribs.

6. Minimum part thickness of about 0.5 mm: smaller values create filling problems resulting from premature solidification.
7. Cycle time that varies from a few seconds to several minutes: the shorter times are for thin products while the longer times are for larger thick-walled products.

1.1.2 The Plasticating Unit

The most common plasticating unit on injection molding machines is a single screw extruder. The extruder is somewhat different from extruders used in continuous extrusion; in these machines the screw rotates continuously but it does not move forward or backward. The screw in an injection-molding machine can rotate but can also move backward and forward. Such an extruder is called a reciprocating screw extruder; other terms used are ram-screw, recipro-screw, or RS extruder.

1.1.2.1 The Extruder Screw

The main components of an extruder are shown in Fig. 1.6.

The most important component is the extruder screw [32]; it is a straight cylinder with one or more helical flights wrapped around it; see Fig. 1.7.

Such a device is often referred to as an Archimedean screw, after Archimedes who developed the basic screw conveyor thousands of years ago. The extruder screw determines the conveying, the heating, the mixing, and in some case the degassing of the plastic. Degassing or devolatilization is done on a vented extruder; these are machine with a vent opening in the barrel through which volatiles can escape. Vented extruders require a special screw geometry, the so-called two-stage extruder screw. The extruder barrel is a straight cylinder.

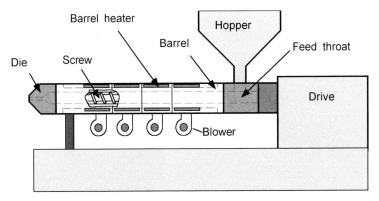

Figure 1.6 The main components of a screw extruder

1 Injection Molding Technology

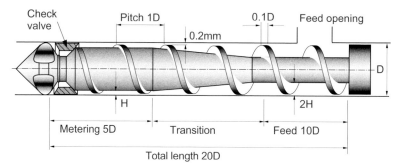

Figure 1.7 A typical extruder screw for injection molding

1.1.2.2   Nonreturn or Check Valves

The screw in a reciprocating extruder usually has a nonreturn valve, also called a check valve, at the end of the screw. The check valve closes when the screw moves forward and opens when the screw moves backward. It keeps the plastic melt from leaking back into the screw when the screw moves forward to inject the melt into the mold. Many different check valves have been developed over the years. Commonly used valves are the ring check valve and the ball check valve.

The action of the ring check valve is illustrated in Fig. 1.8. When the screw rotates, plastic is conveyed forward, melted, and mixed. The plastic melt accumulates at the discharge end of the screw and pushes the screw back against a controlled pressure. As the screw moves backward, the check ring is dragged to the most forward position against a stop at the end of the screw; see Fig. 1.8a. The stop is usually a star shaped shoulder with four to six points. When the check ring rests against the stop, plastic melt can flow through the valve.

When the screw moves forward, the check ring is dragged to the most rearward position against the check ring seat forming a seal; see Fig. 1.8b. In this position the valve is closed and the plastic melt is thus prevented from leaking back into the screw channel during injection. Because of the relative movement between the check ring and the stop, stops will wear over time and eventually have to be replaced.

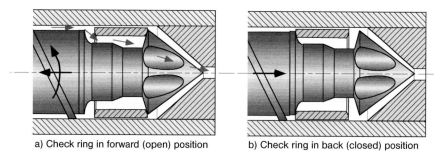

a) Check ring in forward (open) position   b) Check ring in back (closed) position

Figure 1.8 The ring valve in open (a) and closed (b) positions

The operation of a ball check valve is shown in Fig. 1.9. When the screw moves backward, the ball is in the most forward position and the plastic melt can flow to the end of the screw; see Fig. 1.9a. When the screw moves forward, the ball moves to the most rearward position against the ball seat forming a seal; see Fig. 1.9b. In this position the valve is closed. The movement of the ball is due to the pressure difference across the ball. When the screw rotates and moves backward, the pressure behind the ball is higher and the ball is pushed forward. When the screw moves forward, the pressure in front of the ball is higher and the ball is pushed backward.

Some nonreturn valves are designed to mix the plastic melt as it flows through the valve. The advantage of such a valve is that mixing action can be achieved with a simple change to the screw geometry. An example of such a device is the Chris Rauwendaal dispersive (CRD) nonreturn valve, shown in Fig. 1.10.

In some cases a plain screw tip is used without an actual nonreturn valve. The most common application of such a screw is for high viscosity, thermally sensitive materials, such as rigid PVC. Figure 1.11 shows a plain screw tip; it has a small gap between the tip and the barrel. The gap has to be small enough so that only a very small amount of plastic will leak back when the screw moves forward to inject the melt into the mold. The advantages of the plain tip are well-streamlined flow, no hang-up of plastic, and no need for maintenance.

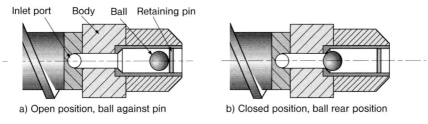

Figure 1.9 A ball check valve in open (a) and closed (b) positions

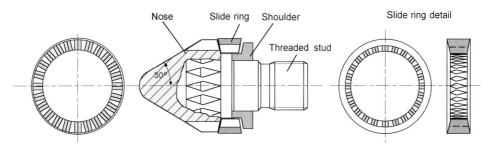

Figure 1.10 A CRD nonreturn valve [41]

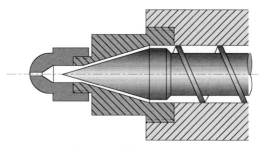

Figure 1.11 A plain screw tip as used for rigid PVC

1.1.2.3 The Extruder Barrel

The extruder barrel is a straight cylinder that closely fits around the screw. The radial clearance between the screw and barrel is typically 0.20 mm (0.008 inch); extruders 40 mm in size and smaller have a clearance of about 0.15 mm (0.006 inch) [1]. The barrel is often manufactured with a bimetallic liner that is centrifugally cast into the barrel. When the plasticating unit is vented, the barrel has a vent opening in the barrel. At the feed end of the extruder, a feed opening is machined in the barrel. This opening is connected to the feed hopper, through which solid plastic particles are introduced to the extruder.

1.1.2.4 The Nozzle of the Plasticating Unit

The extruder nozzle is pushed against the sprue bushing. The nozzle tip usually has a radius that is slightly smaller than the mating radius of the sprue bushing to achieve a good seal. The radius has to be large enough to avoid excessive wear. In the US, standard nozzle radii are 0.5 and 0.75 inch; in Europe, the Euromap standards are 10, 15, 20, and 30 mm. Nozzles can either be open or can contain a shut-off device. Open nozzles are recommended in the processing of thermally sensitive polymers and high viscosity polymers, such as rigid PVC, thermosets, and elastomers. An example of an open nozzle is shown in Fig. 1.12.

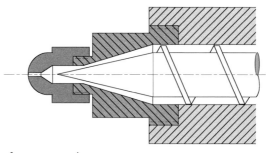

Figure 1.12 Example of an open nozzle

Shut-off nozzles can be used to avoid drooling of molten plastic or stringing; they are also used to be able to have the extruder running with the nozzle retracted. Some nozzles are positively actuated during the cycle; others are controlled separately. An example of a sliding-bolt valve is shown in Fig. 1.13.

Another commonly used nozzle valve is the needle valve shown in Fig. 1.14. When the needle is retracted, as shown in Fig. 1.14a, the valve is open. When the needle is pushed against the nozzle, as shown in Fig. 1.14b, the valve is closed.

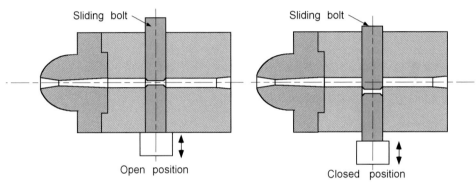

Figure 1.13 Example of a sliding-bolt shut-off nozzle

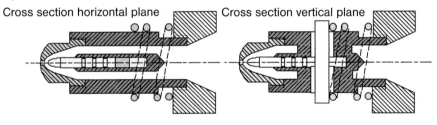

a) Open position, needle retracted

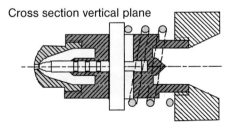

b) Closed position, needle pushed against nozzle

Figure 1.14 Schematics of a needle valve in open (a) and closed (b) positions

1.1.2.5 The Feed Hopper and Feed Opening

The hopper should be designed to achieve a steady flow to the extruder. The hopper cross section is preferably circular to minimize the chance of hang-up of material. The converging region should have a gradual compression to reduce the chance of bridging. Figure 1.15 illustrates the desirable design features in a feed hopper.

Hopper design is critical when dealing with a feed with difficult bulk flow characteristics. Difficult materials are those with a wide range of particle sizes and/or shapes. Bulk materials that are highly compressible tend to cause problems in the hopper region. With such materials special features can be used to ensure steady flow; examples are vibrating pads, low friction coatings, internal cones, rotating wiper arms in the hopper, and others.

It is important that the feed region of the extruder can be maintained at low enough temperature to avoid sticking of the plastic to the metal walls. When plastic particles stick to the metal walls, the size of the flow channel reduces and thus the flow rate reduces. In an extreme case, flow can stop completely. Cooling capability is usually provided in the feed region of the extruder so that sticking of the plastic particles can be avoided.

1.1.2.6 The Extruder Drive

The extruder drive supplies the power to turn the screw and to push the screw forward during injection. Older injection molding machines use hydraulic drives. The hydraulic system can be used to turn the screw, to push the screw forward, and to open and close the mold. More recently, electric molding machines have been introduced. These machines use electric motors to turn the screw, to push the screw forward, and to open and close the mold.

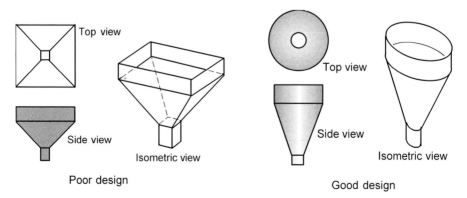

Figure 1.15 Examples of poor and good hopper design

These all-electric molding machines used to be available only in small-to-midrange tonnage categories. As of the year 2000, however, all-electric machines are available up to 1100 tons and plans are in the works for machines up to 2500 tons. There are also a number of electric/hydraulic hybrid molding machines that combine electric with hydraulic drives.

In hydraulic molding machines the hydraulic drive consists of an electric motor and a hydraulic pump. The hydraulic pump converts the energy of the motor into hydraulic energy in the form of pressure. Both constant displacement pumps and variable displacement pumps are used. The pump efficiency should be about 90% at maximum operating pressure; the efficiency of the hydraulic system as a whole will be lower than the pump efficiency.

Hydraulic pumps should be located below the oil level, either at the bottom of the oil reservoir or below it. The drive unit has to be mounted so that noise propagation is minimized; this can be done by rubber support members and hose connections for all the pipelines.

1.2 The Clamping Unit

The main components of the clamping unit are shown in Fig. 1.16.

The main components of the clamping unit are the stationary and moving platens, the tie bars, and the mechanism for opening, closing, and clamping. The mold halves are mounted on the platens; the relative movement between the platens allows the mold halves to open and close. There are three major types of clamping systems: mechanical, hydraulic, and hydromechanical. The clamping force is usually rated in tons, sometimes in kN (kilo-

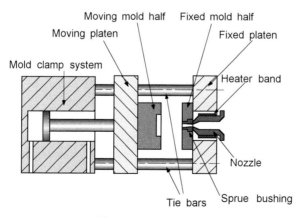

Figure 1.16 The main components of the clamping system

newtons). One metric ton is 1000 kilogram-force; one metric ton equals 9810 newtons. One US ton is 2000 pound-force or about 0.9 metric ton.

For a hydraulic system, the force is determined by the product of pressure and area. A hydraulic clamp with a diameter of 50 cm and a maximum pressure of 10 MPa (1 megapascal = 10^6 pascal; 1 pascal = 1 N/m^2) will give a clamping force of 1,962,500 newtons. The area of a cylinder with 50 cm diameter is $3.14 \times 50^2/4 = 1962.5$ cm^2 = 0.19625 m^2. Multiplying the area with the pressure of 10^7 N/m^2 gives the clamping force of 1962.5 kN or 200 metric tons. Machines range from a clamp force of as little as 5 tons to as much as 10,000 tons.

The clamp force pushes the mold halves together, while the pressure of the plastic melt inside the mold pushes the mold halves apart. The clamping force has to be higher than the separating force from the melt to ensure that the mold halves stay together properly. If this is not the case, the mold can open and melt will flow out at the parting line — this phenomenon is referred to as flash. Obvious ways to reduce flash are by reducing the injection pressure (separating force) or by using a press with a larger clamp force.

The separating force exerted by the melt in the mold can be determined from the melt pressure and the projected area of the mold; this is area projected on the separating plane (parallel to the clamp platen). For instance a compact disk with a diameter of 12 cm will have a projected area of 113 cm^2 (= $3.14 \times 12^2/4$). The melt pressure in the mold can be very high, as high as 2000 bar or 2E8 pascals (29,000 psi). Compare this with the typical air pressure in a car tire of about 2 bar (29 psi).

With a projected area of 113 cm^2 and a melt pressure of 2000 bar = 20,000 N/cm^2, the separating force is $113 \times 20,000 = 2,260,000$ N or 2260 kN; this is about 230 metric tons over 113 cm^2 or about 2 metric tons/cm^2 projected area. This assumes that the melt pressure is transmitted to the entire volume of the mold. In reality this is not the case; there are considerable pressure losses that occur as the melt flows from the mold entrance (sprue) to the far corners and edges of the mold. When a small pinpoint gate is used, the pressure drop is higher than when a large sprue gate is used; gates will be covered later in this chapter. As a result, the clamping force can be considerably less than 2 metric tons/cm^2 projected area without causing opening of the mold. The clamping force per unit projected area is usually in the range of 0.4 to 0.7 metric tons per square centimeter (3 to 5 US tons/inch^2).

1.2.1 Mechanical Clamping Systems

Mechanical clamping is often done with a toggle clamp system with hydraulic actuation for opening and closing. There are single toggle systems and double toggle systems. The single toggle clamp with double acting hydraulic actuator is used in small machines with a clamp force of up to 50 metric tons (about 55 US tons). The actuating cylinder can be connected directly to the toggle with crosshead links or is pivoted at the tailstock platen or the machine support.

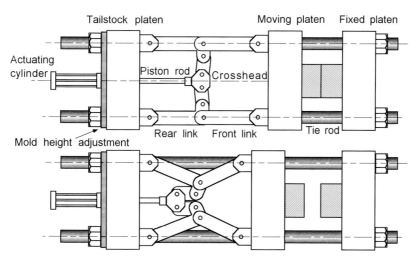

Figure 1.17 Example of a double toggle clamping system

The double toggle clamp is often used on machines with a clamping force between 250 and 1000 metric tons. A double toggle clamp system with central hydraulic actuator is shown in Fig. 1.17. This system uses a central double acting cylinder with five pivot points. Toggle systems usually have a short overall length and save floor space. Compared to hydraulic systems, toggle systems have a relatively short stroke.

Toggle clamping systems provide faster mold opening and closing than hydraulic clamping systems; they are also less expensive than hydraulic systems, at least for smaller injection molding machines. Another advantage of toggle systems is that they can be self-locking if they are designed to go "over center." Once the links reach their extended position, they will remain there until retracted. On the down side, toggle systems have more parts that can wear out as compared to a hydraulic system. There is no simple indication of the clamping force, and the clamping may not remain constant.

1.2.2 Hydraulic Clamping Systems

In hydraulic systems a hydraulic cylinder and piston are used to open and close the mold. These systems are used on most large injection molding machines. Advantages of hydraulic clamping systems are:

- The clamping force can be controlled.
- The clamping force can be easily monitored.
- The ram speed can be controlled.
- The stroke of the ram can be adjusted easily.

- Low-pressure protection is easy to incorporate.
- The breakaway opening force and speed are adjustable.
- Low maintenance.
- They allow fast mold setup.

The major disadvantages of a hydraulic clamping system are the higher initial cost, higher operating cost, and the chance of oil leakage.

1.2.3 Hydromechanical Clamping Systems

A hydromechanical system uses both toggle and hydraulic systems together. Thus these systems can combine some of the advantages of both systems, such as highspeed open and close and precise control of clamping force. In hydromechanical systems, a piston or other device provides fast initial closing, while a hydraulic system is used in the final clamp stroke. The precision of the hydraulic phase eliminates the danger of overtoggling.

1.3 The Mold

The mold typically consists of two mold halves; see Fig. 1.18. Usually one mold half contains the cavity and forms the outer shape of the part; appropriately, this part is called the cavity. The other mold half contains a protruding shape and forms the inner shape of the part; this mold part is called the core. When the core is clamped against the cavity, the hollow space that is formed defines the shape of the part to be molded. The plastic is usually injected into the mold from the cavity side. For that reason the cavity side is also called the "hot side" or the "hot half." The cavity side is typically mounted on the stationary platen and, therefore, is also referred to as the stationary half. The core side is typically mounted on the moving platen; the core side, therefore, is often called the moving half. The ejection mechanism is often part of the moving platen. The product is usually removed or stripped from the core side.

Parts and cavities have to be designed such that demolding is possible after the mold opens at the parting line and the ejectors are activated. The molded part usually stays in the movable mold half when the mold opens. When the ejector or stripper mechanism is activated, the molded part(s) is pushed out of the cavity or off the core. When the mold closes, the ejector mechanism returns to its original position.

1.3 The Mold

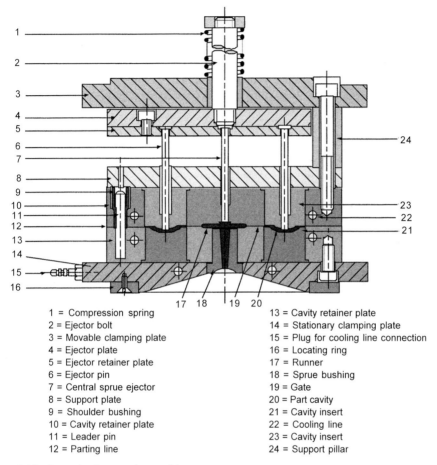

1 = Compression spring
2 = Ejector bolt
3 = Movable clamping plate
4 = Ejector plate
5 = Ejector retainer plate
6 = Ejector pin
7 = Central sprue ejector
8 = Support plate
9 = Shoulder bushing
10 = Cavity retainer plate
11 = Leader pin
12 = Parting line
13 = Cavity retainer plate
14 = Stationary clamping plate
15 = Plug for cooling line connection
16 = Locating ring
17 = Runner
18 = Sprue bushing
19 = Gate
20 = Part cavity
21 = Cavity insert
22 = Cooling line
23 = Cavity insert
24 = Support pillar

Figure 1.18 Example of a two-plate mold

1.3.1 The Sprue

The sprue is typically the entry channel of the mold; it is adjacent to the injector nozzle. The sprue is usually a conical protrusion on the runner, as shown in Fig. 1.19. The molten plastic flows through the nozzle into the sprue, then into the runner, from the main runner into any possible secondary runners, then into the gates, and from the gates into the molded part cavities.

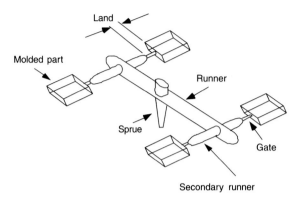

Figure 1.19 A runner system for a four-cavity mold

1.3.2 The Runner System

Once the plastic melt enters the mold, it flows through a distribution system, called the runner system, and then through the gates into the part cavities. There are two main types of runner systems: cold runner systems and hot runner systems.

1.3.2.1 Cold Runner Systems

In a cold runner system, a new runner is molded each molding cycle and the runner is ejected together with the molded part(s). The plastic of the runner can often be reprocessed and molded again. Cold runner systems used to be employed on virtually all injection molding machines. Nowadays, however, hot runner systems are used in a number of applications.

The two main types of cold runner molds are two-plate and three-plate molds. In a two-plate mold the plastic flows to the gate from the parting line and enters the mold cavity at the side of the product; see Fig. 1.20.

In a three-plate mold the runner usually meets with the gate at the bottom at the center of the product; see Fig. 1.21. A two-plate mold may actually consist of more than two plates and a three-plate mold may have more than three plates; despite this, the terms two- and three-plate molds are still maintained.

As can be seen in Fig 1.21, in a three-plate mold the gate is removed from the molded part when the mold opens. Also, the three-plate mold has two parting lines as opposed to one parting line in a two-plate mold. When the mold opens, the cavity and core halves stay together for the first part of the opening stroke. The pinpoint gate, where the drop connects to the cavity, breaks as soon as the cavity plate separates from the runner plate. As the core moves far enough from the cavity the parts can be ejected; also the runner can be ejected as the cavity plate moves far enough from the runner plate.

1.3 The Mold   17

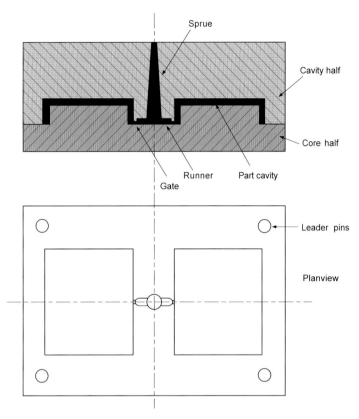

Figure 1.20 Example of a two-plate mold

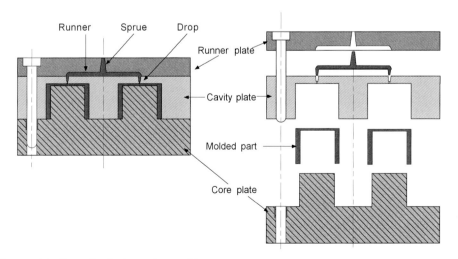

Figure 1.21 Example of a three-plate mold

1.3.2.2 Balancing the Runner System

In the design of the runner system the objective is to have the plastic reach all gates at the same time. This is an important issue in multicavity molds. In a rectangular runner system, the number of cavities is a multiple of two in most cases: two, four, six, eight, and so on. In a circular runner any number of cavities can be used, typically three but higher numbers have been used as well.

In some molds two or more parts are made simultaneously; these molds are often referred to as "family molds." In such a mold the cycle time is determined by the part requiring the longest cooling time, that is, the thickest part. This approach is used in low production items where it can be cost effective to mold several parts together in one mold, thus ensuring that the quantities required of each part are correct. Geometric balancing is not possible in this case, and some fine-tuning of the runner system may be required after mold tryout.

Family molds are used in some high production items when the products have about the same size and molding speed. This can be particularly useful when two matching products are molded that are assembled just downstream of the molding operation. An additional advantage is this minimizes the chance of problems with color matching of the parts assembled together.

Some typical runner systems are shown in Fig. 1.22. There are no major differences in layout of cold and hot runner systems. In cold runner molds the runners can be milled, and thus can be easily made with curves. In hot runner molds the runners are usually drilled and therefore cannot have curves.

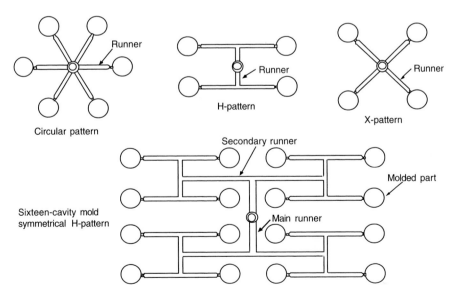

Figure 1.22 Several layouts of runner systems

1.3.2.3 Cold Slug Traps

Cold runner molds may have cold slug traps, which are extensions of each branch beyond the point where the branch occurs; see Fig. 1.23.

These traps are supposed to catch the more forward portion of the plastic flowing into the runner system, which may have cooled down too much and block the flow. This may be an issue when the injection speed is slow; however, with high injection speed these cold slug traps are usually not necessary.

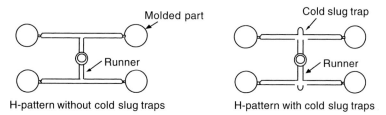

Figure 1.23 Two H-pattern runner systems with and without cold slug traps

1.3.2.4 Gates

The gate connects the runner to the actual part; more than one gate may be used for one part. Gates can be center gates or edge gates. The cross section of the gate is usually small so that the runner can be easily removed from the part and does not leave a large gate mark on the part. Various types of gates can be used: sprue gates, pinpoint gates, edge gates, and tunnel gates, see Fig. 1.24.

The sprue gates are the oldest type of center gate; they are used with very large products. The gate is cut after the part has been removed, leaving a mark or vestige on the product. Sprue gates can often be converted to hot sprue gates; these leave hardly any mark and do not require cutting.

Pinpoint gates are commonly used in three-plate molds. They leave only a small gate mark and are self-degating.

Edge gating uses an entrance channel at the edge of the part cavity. In some cases the entrance channel is made wide and very shallow; such a gate is referred to as a fan gate. A fan gate might be 20 mm wide and 0.12 mm high. The mark left by a fan gate looks like a thin line.

The tunnel gate, also called submarine gate, is used with two-plate molds. During the opening of the mold the gate and runner are retained by undercuts. The runner pulls away from the part at the tunnel and separates from the part. This type of mold, therefore, is degating. Other types of gates are diaphragm gates and ring gates.

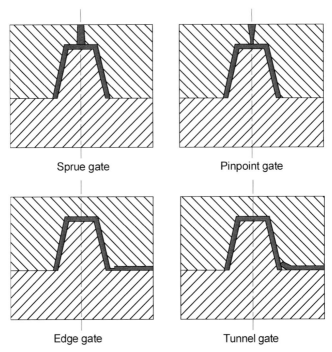

Figure 1.24 Various types of gates

1.3.2.5 Hot Runner Systems

In hot runner systems the plastic is kept hot from the nozzle to the gate. The temperature in the hot runner is usually about the same as the melt temperature in the nozzle. The hot runner normally does not actually heat the plastic; it keeps the plastic from cooling down so that it can stay at the required melt temperature. There are two types of hot runner molds: insulated runner molds and true hot runner molds.

a) Insulated Hot Runner Molds

An insulated hot runner mold is illustrated in Fig. 1.25.

An important requirement for these molds is good access to the runner space between the runner plate and the cavity plate in case the plastic solidifies in the runner. When this happens, the runner plate has to be separated from the cavity plate and the plastic has to be removed from the runner. The chance of the plastic freezing in the runner can be much reduced by incorporating a heater inside the drop. This complicates the design of the mold; on the other hand, it allows molding of plastics that are very difficult to run on regular insulated runner molds, for example PS.

1.3 The Mold   21

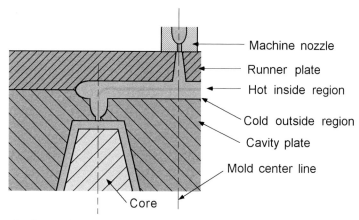

Figure 1.25 Example of an insulated runner mold

b) True Hot Runner Molds

An example of an early version of a hot runner mold is shown in Fig. 1.26.

A hot runner mold is a three-plate mold where the runner plate is heated, usually by cartridge heaters. The hot runner system is basically a continuation of the hot channel of

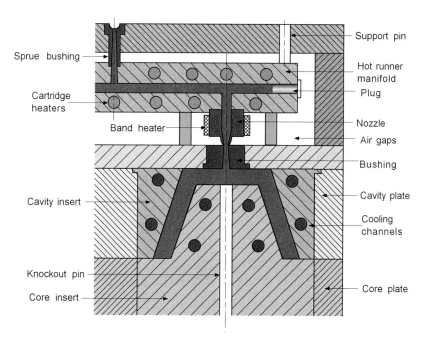

Figure 1.26 Example of a hot runner mold

the extruder nozzle system. One of the challenges with hot runner molds is to keep the nozzles leading to the cavities from freezing. If the nozzle is too hot, the plastic melt will drool, and if it is too cold, the plastic melt will freeze.

Hot Runner Nozzles

In addition to the extruder nozzle, hot runner molds have nozzles at the ends of the hot runner system. These nozzles connect the hot runner manifold to the gates at the entrance to the cavities; see Fig. 1.26. The nozzle should have sufficient mechanical strength to withstand the high internal pressure of the plastic melt and good thermal conductivity to keep the plastic hot. Also, the nozzle should be thermally isolated from the cold parts of the mold to prevent excessive heat loss from the nozzle.

Hot Runner Gates

There are two types of gates, open gates and valved gates. The open gate is the simplest gate, however, it is difficult to control. The open gates should provide unrestricted flow during injection, but freeze off during the cooling part of the cycle. If the open gate does not freeze off during cooling, the melt from the runner system can ooze out of the hot runner.

Valved gates provide some type of mechanical control of the flow through the gate; numerous designs of valved gates are available. In valve gates, typically a movable pin reaches into the gate. In the forward position of the pin the gate is closed; in the rearward position the gate is open. An example of a spring-loaded valved gate is shown in Fig. 1.27, left side.

A drawback of the spring-loaded valved gate is that there is no direct control over the force on the pin. This drawback is eliminated in the air-actuated gate shown in Fig. 1.27, on the right. This valved gate uses an air cylinder and piston; by varying the air pressure, the force on the pin can be controlled. Some systems use hydraulics to control the motion of the valve pins; however, there are some hazards associated with using oil close to electric heaters. Externally operated cams or levers can also be used to actuate valve pins.

1.3.3 Flow into the Mold

The flow into the mold cavity is normally a flow with an advancing flow front. The cavity fills from the gate with the fountain-shaped flow front advancing to the opposite end of the cavity. This type of flow is referred to as fountain flow; it is illustrated in Fig. 1.28.

The different positions of the advancing flow front are numbered in Fig. 1.28. Elements of the fluid in the center region first decelerate as they approach the flow front, then the elements start to move tangentially towards the wall. The wall is relatively cold and a frozen skin layer will form behind the advancing flow front. The elements in the flow front are stretched as they move from the center towards the wall; this is illustrated in Fig. 1.28.

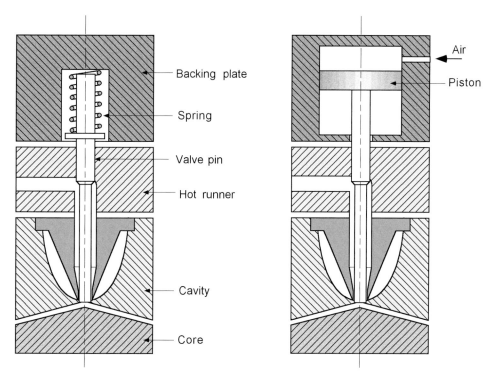

Figure 1.27 Examples of spring-loaded valved gate (left) and air-actuated valved gate (right)

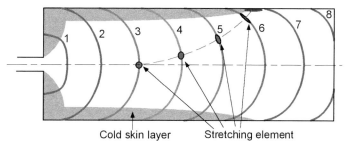

Figure 1.28 Illustration of fountain flow into a mold cavity

The fountain flow mechanism is responsible for a high degree of orientation in the surface layers of the molded product. This is due to the stretching of the fluid elements approaching the wall and the rapid cooling that occurs at the wall. The cooling away from the wall is much slower and, as a result, more relaxation can occur in the internal material elements. Many studies have confirmed significant gradients in orientation and morphology from the outside layers to the inside of injection molded parts.

In some molds the normal fountain flow process does not occur, but the melt emerging from the gate shoots out and collects at the wall opposite to the cavity entrance. This is called jetting; see Fig. 1.29.

When jetting occurs, there is a nonuniform distribution of the plastic melt in the mold, causing poor part appearance and physical properties. The occurrence of jetting is strongly determined by the design of the gate. Various design options are available to reduce the chance of jetting, such as enlarging the gate, using an eccentric gate, and so on. Detailed mold design issues are treated in specialized texts, such as the book by Menges and Mohren, [36] or the book edited by Stoeckhert and later by Mennig [37].

Another problem that is related to the design of the gate is blush; this is a surface condition caused by high stresses occurring in the filling process. It has a streaky behavior that can create a starburst look around a gate. Blush is likely to occur when the gate is machined only into one mold half; see Fig. 1.30.

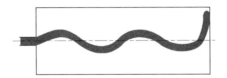

Figure 1.29   Example of jetting

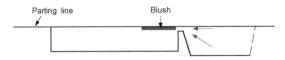

Figure 1.30   Gate machined only into one mold half

1.3.3.1   Venting

From the description of the mold filling process it is clear that the air in the mold will be displaced by the advancing melt front. It is important that the mold is designed in such a way that the air displaced in the mold filling process has a chance to escape from the mold. If air does not have a chance to escape, it is compressed quite rapidly. As the air compresses, it heats up; the temperature rise can be high enough to cause burning of the plastic. Thus improper venting can not only cause incomplete filling of the molded part, but can also cause burn marks.

Venting at the parting line can be done quite easily. In some cases the cavity is surrounded by one or more collector grooves into which the spot vents let the air escape. In other designs, a continuous vent is provided surrounding the cavity. The gap of an air vent must be sized so that the air can escape but not the plastic. The vent gap is usually around 0.01 to

0.02 mm (0.0004 to 0.0008 inch). Obviously, the appropriate dimensions of the vent gap will depend on the viscosity of the plastic. The width of spot vents can vary with the specifics of the mold design; typical values are around 5 to 6 mm (0.20 to 0.24 inch). The land length is usually around 0.6 to 1.2 mm (0.024 to 0.048 inch).

In some molds the plastic melt can flow faster in some directions than others; this can be due to ribs or large cross sections in the cavity. This situation is more common in molds with edge gating than in molds with center gating. When this happens, separate plastic melt streams can join, trapping some air that cannot escape through the vents at the parting line. Ejector pins in the area of such traps can act as vents, these are called natural vents. If ejector pins are not present, vent pins or inserts with vents have to be used to ensure proper venting.

1.3.3.2 Weld Lines

When the plastic melt flows around an obstacle in the mold, the melt will recombine after the obstacle and a weak region will tend to form where the melt streams meet; see Fig. 1.31. This is a common problem not only in injection molding, but also in extrusion of hollow products. The meeting of separate melt streams also occurs in molds with multiple gates. The weak region that forms is commonly called a weld line, even though it is not really a line, but more of a plane.

The weakness is due to the fact that the plastic molecules take a certain amount of time to reentangle after melt streams meet. This reentanglement is dependent on the melt temperature and the molecular weight of the plastic. The lower the melt temperature and the higher the molecular weight, the longer the reentanglement process takes. If the molding time is less than the time to reach full reentanglement, the molded part will have a

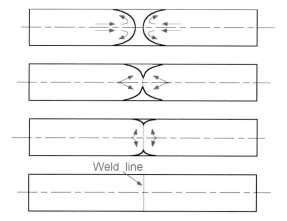

Figure 1.31 Formation of a weld line

weak region. When the molded part is exposed to stresses, it is more likely to fail at the weld line than in other regions of the part.

1.3.3.3 Mold Filling Analysis

The properties of the molded product are to a large extent determined by the flow process into the mold. Problems with weld lines, blush, and so on, are very much determined by the design of the mold, the location of the gate(s), the design of the gate(s), and so on. The filling process is clearly a complicated process; a few useful tools are available to aid the analysis of mold filling.

An experimental method to follow the filling process is to produce short shots. The cavity is intentionally filled with an insufficient amount of material to show how the cavity fills; for instance, jetting can be firmly identified with short shots. A theoretical method to analyze mold filling is the use of computer simulation software. Currently, there are a number of commercial packages that allow the user to simulate the mold filling process. These programs can predict filling patterns, temperature profiles, shear stresses in the mold, and so on. This information can be valuable in the design of the mold and in troubleshooting molding problems.

Newer and improved versions of mold filling software not only simulate the filling process, but also the cooling of the material in the mold and the deformation of the molded part after it leaves the mold. Thus one can perform a warpage analysis based on computer predictions. Using such software requires a reasonable level of skill; as a result, these programs tend to be used more in large companies than in small molding operations. Discussion of theory of mold filling and the numerical techniques is beyond the scope of this book; the book by P. Kennedy, "Flow Analysis of Injection Molds" [38], deals with this subject in detail.

2   Extrusion Technology

2.1   Introduction

Extrusion is a process for the continuous production of plastic products such as tubing, pipe, film, sheet, coated wire, fiber, profiles, and others. Most extruded products have a constant cross section; however, it is also possible to make products with a nonconstant cross section. An example of the latter is extrusion blow molding, where a parison is extruded and, subsequently, inflated with air in a mold to make a blow molded product. Another example is bump tubing used in medical applications. In this tubing, sections of larger diameter are produced next to smaller diameter sections in a reproducible and repeatable fashion.

Injection molding is a process for discontinuous production of plastic parts. Injection molded parts usually do not have a constant cross section. Generally, parts with constant cross section can be made more economically by extrusion than by injection molding. Even though there are major differences between extrusion and injection molding, there are also many similarities. Both processes use an extruder to convey, melt, and pressurize the plastic. In a regular extruder the screw only rotates, whereas in an injection molding extruder the screw generally rotates and moves axially at the same time. Many process improvements that can be made to extrusion can also be used in injection molding. For instance, changes in screw design that improve extruder performance generally can be used beneficially in injection molding as well.

There are various aspects of the machine design and the processing method that strongly affect the stability of the process and the quality of extruded or molded product. The resulting variability, in most cases, is ever present and is, in a statistical sense, predictable. These causes of variation, therefore, are common or inherent causes. Statistical process control (SPC) focuses primarily on special or unnatural causes. However, it is very important not to stop when special causes of variation have been eliminated. Unfortunately, SPC offers little guidance in the elimination of common cause variations. The most important ingredient in eliminating common cause variation is a good knowledge and understanding of the process technology. SPC is a very useful tool in process analysis, improvement, and control. However, SPC by itself is not enough; it needs to be combined with process technology knowhow to really pay dividends.

The type of extruder most commonly used is a single screw extruder. A typical single screw extruder is shown in Fig. 2.1. The main components of an extruder are:

1. The extruder screw
2. The extruder barrel
3. The feed hopper

28 2 Extrusion Technology

4. The feed throat casting
5. The die
6. The drive for the screw
7. Heating and cooling elements
8. Instrumentation and control system

The extruder screw is a long cylinder with one or more helical flights wrapped around it. Figure 2.2 shows a simple conveying screw. The screw is the heart of the extruder. It is very important because the conveying, heating, melting, and mixing of the plastic are mostly determined by the screw. As a result, the stability of the process and the quality of the extruded product are very much dependent on the design of the screw.

Simple conveying screws have poor mixing capability. Therefore, mixing sections are often incorporated in the screw design to improve mixing performance. There are many different mixing sections [42, 43]. Most mixing sections are either distributive mixers or dispersive

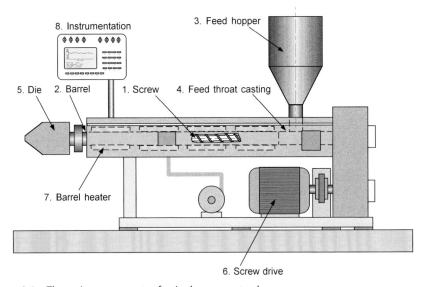

Figure 2.1 The main components of a single screw extruder

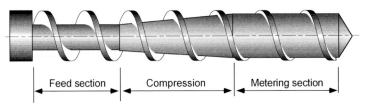

Figure 2.2 Simple conveying screw, single flighted and single stage

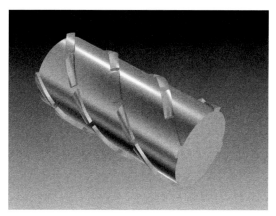

Figure 2.3 CRD mixing section

mixers. One mixer that is both a distributive and dispersive mixer is the CRD mixer [44], shown in Fig. 2.3.

The screw is located in the extruder barrel. The barrel is a long cylinder with a small clearance between the screw and the barrel. A typical radial clearance is onethousandth of the barrel diameter. The screw normally rotates in a stationary barrel. The rotation of the screw causes a forward conveying of the plastic from the feed opening to the die exit. When the extruder is fed with solid plastic particles, it is called a plasticating extruder. When it is fed with molten plastic, it is called a melt fed extruder. We will deal mostly with plasticating extruders since these are most common.

2.2 The Functions of an Extruder

The main functions performed in an extruder are:

1. Conveying
2. Heating and melting
3. Mixing
4. Die forming
5. Degassing (in a vented extruder)

2.2.1 Conveying

The forward conveying is due to drag forces between the plastic and the barrel. The drag forces between the plastic and the screw actually work against the forward conveying. Without a drag force against the barrel, there is no forward conveying, only rotation with the screw. For good, steady conveying it is important, therefore, to have a large drag force at the barrel and a small drag force against the screw. This is why grooves in the barrel surface at the feed end of the extruder can dramatically improve the conveying characteristics of an extruder.

The conveying in the extruder barrel is due to the rotation of the screw. The flow resulting from screw rotation is called drag flow. The conveying through the extruder die is due to the pressure developed by the screw. The flow through the die is due to the difference in pressure between the inlet and outlet of the die. Since the outlet pressure is constant at atmospheric pressure, the extruder output is determined directly by the die inlet or diehead pressure. Thus the screw is the pressure-generating element, while the die is the pressure-consuming element. The extruder output, thus, is determined by the interaction of the extruder and the die.

The output of the extruder decreases as the pressure at the end of the screw increases. This is shown by the screw characteristic curve in Fig. 2.4. On the other hand, the output from the die increases as the inlet pressure to the die increases. This is shown by the die characteristic curve in Fig. 2.4. Thus each extruder and each die have their own characteristic curves. These are determined by the machine geometry, operating conditions, such as temperature and screw speed, and the flow properties of the plastic. The actual output and head pressure of an extruder/die combination is determined by the intersection of the extruder and die characteristic curves; this is the operating point as shown in Fig. 2.4.

The die characteristic curve is primarily determined by the geometry of the die flow channel, the die temperature, and the plastic melt viscosity. A die with very thin, restrictive, flow passages will require high pressures to obtain reasonable flow rates; see Fig. 2.5. A die with large flow passages will require only low pressures to obtain reasonable flow rates, see

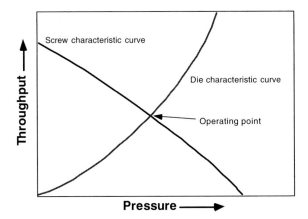

Figure 2.4 Screw and die characteristic curves

Fig. 2.5. Temperature also affects the die characteristic curve. A high die temperature reduces the restriction to flow and results in a lower diehead pressure at the same output or a higher output at the same diehead pressure. The lower die flow restriction at higher temperature is due to the fact that the plastic melt viscosity reduces with increasing temperature. In other words, the plastic melt flows more easily at higher temperatures.

The extruder characteristic curve is primarily determined by screw geometry, the screw speed, barrel temperatures, and the flow properties of the plastic. A screw with deep channels will give high output at low head pressure, but the output will drop quickly with pressure. A screw with shallow channels will give low output at low head pressure, but the output will not drop quickly with pressure; see Fig. 2.6. Therefore, a high restriction die should be combined with a shallow screw, while a low restriction die should be combined with a deep screw; see Fig. 2.6.

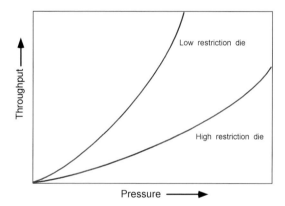

Figure 2.5   Die characteristic curves for low and high restriction dies

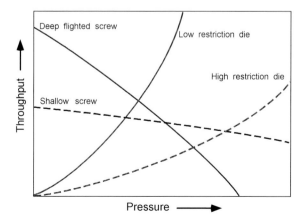

Figure 2.6   Characteristic curves of shallow/deep screws and low/high restriction dies

2.2.2 Heating and Melting

There are two sources of heat in the extrusion process. One is the heat from heating elements placed along the extruder; the other is the frictional and viscous heat generation within the plastic. Frictional heating occurs when solid plastic particles slide past a metal surface or past one another. Viscous heating occurs when the plastic melt is sheared by the rotation of the screw. The viscous heating is primarily determined by the shear rate in the melt and the viscosity of the melt. The shear rate is a measure of the shearing action in a fluid. It is determined by the difference in velocity between two points divided by the distance between those points; see Fig. 2.7. In an extruder the approximate shear rate in the screw channel is the circumferential velocity of the screw divided by the channel depth.

As an example, let's consider a 150 mm (6 inch) extruder running at 90 rpm and a screw channel depth of 7 mm (0.28 inch). The circumference of the screw is 3.14 × 150 = 471.24 mm. The circumferential speed is 471.24 × 1.5 = 706.86 mm/s. The 1.5 comes from 90 rev/min = 1.5 rev/s. The approximate shear rate becomes 706.86/7 = 100.98 (1/s). The shear rate is expressed in 1/s or s^{-1}; this is called reciprocal seconds. Typical shear rates in a screw channel range from 50 to 100 (s^{-1}). The shear rates in an extruder increase with screw speed.

As a result of the high plastic melt viscosity, the viscous heat generation is often substantial. In fact, in many extruders most of the heat comes from frictional and viscous heat generation. This is particularly true at high screw speed. At low screw speed the contribution from the barrel heaters is large and the viscous heat generation relatively low. However, as the screw speed increases, the contribution from viscous heating increases and the contribution from the barrel heaters reduces. This is illustrated in Fig. 2.8. The barrel heaters usually contribute only a small amount of heat to the extrusion process. The frictional and viscous heat generation is essentially a transformation of mechanical energy from the extruder drive into thermal energy to increase the plastic temperature.

When the plastic temperature reaches its melting point, melting will start. Melting usually starts at the barrel surface about five diameters from the feed opening.

Initially, a melt film forms on the barrel surface. As the melt film grows thicker, a melt pool forms at the leading flank of the flight, pushing the solid bed against the trailing flank of the flight. This model, first described by Maddock [47], is shown in Fig. 2.9.

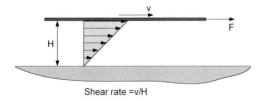

Figure 2.7 Illustration of shear rate

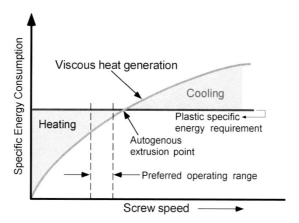

Figure 2.8 Specific energy consumption versus screw speed

The two main sources of heat for melting are the heat from the barrel heaters and the viscous heat generation in the melt film between the solid bed and the barrel. For a high melting efficiency, it is important to keep the melt film as thin as possible. The melt film thickness is strongly determined by the screw flight clearance. Therefore, the flight clearance should be kept small. If the flight clearance increases, for instance as a result of screw wear, the melting rate can reduce substantially.

The melting model shown in Fig. 2.9 is also called the contiguous solids melting (CSM) model. It has been observed in many experimental studies of melting in single screw extruders. In some cases, however, another melting is observed where the solid particles are discrete and floating in a melt matrix. This type of melting is called dispersed solids melting (DSM) [45] or dissipative melt mixing [46], see Fig. 2.10. DSM has been observed in twin screw extruders and reciprocating single screw compounders. Melting by DSM occurs more efficiently than by CSM. In twin screw extruders, the melting length is often only about two to three diameters long, while in single screw extruders the melting length is typically around ten to fifteen diameters long.

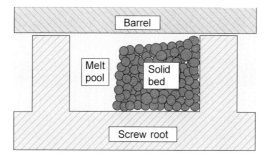

Figure 2.9 Maddock's melting model

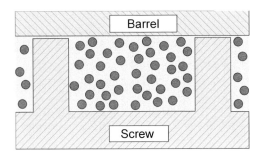

Figure 2.10 Dispersed solids melting model

When the barrel temperature is increased, the amount of heat for melting from the barrel heaters increases. At the same time, however, the viscous heat generation in the melt film will reduce because a higher barrel temperature will reduce the viscosity in the melt film. At low speed, when the barrel heaters supply a major portion of the heat, higher barrel temperatures will increase melting; see Fig. 2.11. At high screw speed, however, the major portion of the heat for melting comes from viscous heat generation. At high speed, therefore, increased barrel temperature can actually reduce melting in the extruder; see Fig. 2.11.

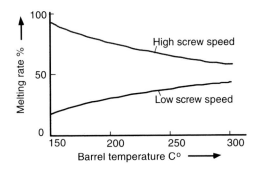

Figure 2.11 Effect of barrel temperature on melting rate

2.2.3 Mixing

Extruder screws without special mixing devices have relatively poor mixing capability. There are several reasons for this. Material that melts early will be exposed to a long shearing action and will be reasonably well mixed. However, material that melts late will be sheared only for a short time and will not be well mixed. Also, because of the recirculating flow across the screw channel, demixing effects occur in the screw channel. This is due to the fact that the direction of shearing in the bottom section of the screw channel is opposite to the shearing in the upper portion of the channel. This is illustrated in Fig. 2.12.

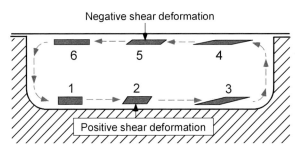

Figure 2.12 Demixing effects in the recirculating flow in the screw channel

The recirculating flow also causes nonuniform temperatures, as shown in Fig. 2.13. Heat transfer occurs primarily at the barrel surface. As a result, the temperature of the material in the outer recirculating flow will stay relatively close to the barrel temperature. The material in the inner recirculating flow, however, will be insulated from the barrel. Therefore, its temperature can rise substantially above the barrel temperature, by as much as 50 °C (90 °F) or more. The heat transfer depends strongly on the flight clearance. A large flight clearance reduces the recirculating flow, and thus even higher melt temperatures can occur over a larger area of the screw channel.

Because of the inherent nonuniform mixing and melt temperatures in a standard extruder screw, it is important to incorporate efficient mixing devices to improve the melt quality. Mixing is usually divided into distributive mixing and dispersive mixing.

Distributive mixing is the mixing of the fluids with similar flow properties. Dispersive mixing is the mixing of a fluid with a solid filler or the mixing of two incompatible fluids. The agglomerates of the filler or liquid droplets have to be broken down by the stresses that the fluid exerts on the filler or droplets.

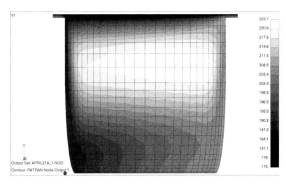

Figure 2.13 Melt temperature distribution in screw channel; barrel temperature is 175 °C

2.2.3.1 Distributive Mixing

Requirements for efficient distributive mixing are:

- Large shear strain
- Frequent flow splitting
- Frequent reorientation

Shear strain is the product of shear rate and shear exposure time. Thus a shear rate of 100 (1/s) for 20 seconds will give a shear strain of 2000. A lower shear rate of 10 (1/s) will require a longer shear exposure time of 200 seconds to achieve the same shear strain. A large shear strain by itself does not necessarily give good mixing. Splitting and reorientation are very important in increasing the mixing efficiency substantially. Table 2.1 shows a comparison of commercial distributive mixers. The ranking is based on a five-point system: 5 is excellent, 1 is very bad.

Table 2.1 Comparison of Various Distributive Mixers

Mixers	Pressure drop	Dead spots	Barrel wiped	Operator friendly	Machining cost	Shear strain	Splitting and reorientation
Pins	2	2	3	4	5	2	4
Dulmage	4	4	2	4	4	4	5
Saxton	4	4	5	4	4	4	5
CTM	1	3	2	1	1	4	5
TMR	1	3	4	3	3	4	5
Axon	4	4	4	4	5	4	3
Double wave	4	4	4	4	2	4	2
Pulsar	4	4	4	4	3	3	2
Stratablend	4	3	4	4	3	3	2

2.2.3.2 Dispersive Mixing

Dispersive mixing is more difficult to achieve than distributive mixing. Single screw extruders generally have poor dispersive mixing capability, while twin screw extruders tend to have much better dispersive mixing capability. Dispersive mixing requires the application of high stresses to break down agglomerates or droplets. It has been found that elongational stresses are much more effective in achieving dispersion than shear stresses [43]. Conventional dispersive mixers for single screw extruders are based on shear flow, while twin screw dispersive mixers rely on elongation flow. This explains why twin screw extruders usually do a better job in dispersive mixing than single screw extruders.

Recently, a mixer (CRD) was developed for single screw extruders [44] that relies on elongational flow; see Fig. 2.3. This mixer uses the same mixing mechanism as twin screw extruders and, as a result, its dispersive mixing capability is comparable to that of twin screw extruders. Table 2.2 shows a comparison of commercial dispersive mixers for single screw extruders. The ranking is based on a five-point system: 5 is excellent, 1 is very bad.

Table 2.2 Comparison of Dispersive Mixers for Single Screw Extruders

Mixer	Pressure drop	Type flow	Dead spots	Barrel wiped	User friendly	Machining cost	Number passes
Blister ring	1	Shear	3	2	4	5	1
Egan mixer	2	Shear	5	5	4	4	1
LeRoy (Maddock)	2	Shear	2	5	4	4	1
Zorro mixer	5	Shear	5	5	4	3	1
Helical LeRoy	5	Shear	5	5	4	4	1
Planetary gear	3	Shear	5	5	4	2	> 1
CRD mixer	5	Elongation	5	5	4	4	> 1

Another important issue in the performance of dispersive mixers is how many times the fluid elements pass through the high stress region. A fine level of dispersion generally requires that all fluid elements pass through the high stress region several times. It is interesting to note that in most conventional dispersive mixers the material passes through the high stress region only once. Of the mixers listed in Table 2.2, only the planetary gear and CRD (Chris Rauwendaal Dispersive) mixers achieve multiple passes through the high stress region.

2.2.4 Die Forming

A wide variety of products can be made by extrusion: tubing, pipe, bags, netting, fiber, film, sheet, bottles, fuel tanks, window profiles, and so on. For products with a constant cross section, the shape of the extruded product is primarily determined by the geometry of the flow channel in the die. For other products, considerable shaping can occur beyond the die. Examples are corrugated tubing, bump tubing, blow molded products such as bottles, thermoformed products such as pickup truck bed liners, and so on. In fact, injection molding can be considered a cyclic extrusion process. Proper design of the extrusion die is very important.

The flow through the die results from the pressure difference between the inlet and outlet of the die. A number of difficulties can occur in the flow through dies. The restriction to flow can vary along different streamlines in the die. This will result in velocity differences and

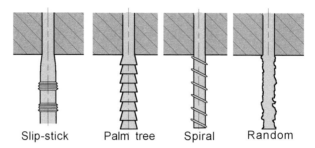

Figure 2.14 Various forms of melt fracture

distortion of the extruded product. Excessive stresses may occur in the die; this can lead to flow instabilities. An example is melt fracture; this causes a severe distortion of the extruded product; see Fig. 2.14.

When the plastic melt leaves the die, it tends to swell. Unfortunately, this swelling is usually nonuniform and distorts the shape of the product. The take-up speed is usually higher than the extrusion speed to maintain some tension in the line. This causes a drawdown of the extruded product and changes its size and shape, often resulting in an undesired product geometry.

If the extruder screw does not deliver a uniform plastic melt to the die, uneven flow can occur in the die resulting in varying product dimensions. The melt flowing into the die should be uniform in temperature and consistency. Therefore, the extruder screw should have good mixing capability. Uneven melt temperatures can also result in uneven stresses in the product. This can lead to distortion of the product later when the stresses relax.

In hollow products, such as tubing and pipe, the plastic melt has to flow around a central core or mandrel. The mandrel is usually held in place by a number of spider supports; see Fig. 2.15.

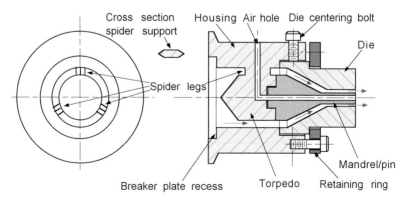

Figure 2.15 In-line tubing or pipe die

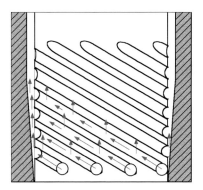

Figure 2.16   A spiral mandrel die

As the plastic melt flows past the spider supports, it splits and flows together again after the spider. When the melt stream recombines, it takes a certain amount of time before full, intimate contact is established. This is because the long plastic molecules take some time to reentangle. If insufficient time is available for the molecules to reentangle, a weak spot will form along the length of the extruded product. This is called a weld line or knit line. Such a weld line can cause premature failure of a product, particularly if the product is internally pressurized (which is often the case, e.g., in tubing and pipe). A design for annular products that minimizes the weld line problems is the spiral mandrel die, see Fig. 2.16.

2.2.4.1   Guidelines for Shapes

In order to make a product easy to extrude, the following guidelines should be used regarding the shape of the product:

- Use generous radii at corners (larger than 1 mm).
- Use uniform thickness for exterior walls.
- Avoid very thick walls (larger than 10 mm).
- Minimize the use of hollow sections.
- Make interior walls thinner (about 30%) than exterior walls.
- Make products symmetrical if possible.

2.2.4.2   Guidelines for Die Design

The following guidelines can be useful in selecting a die:

- The flow channel geometry should allow streamlined flow without abrupt transitions or dead spots.
- The cross-sectional area of the flow channel should reduce gradually from the inlet to the outlet to achieve a gradual increase in flow velocity.

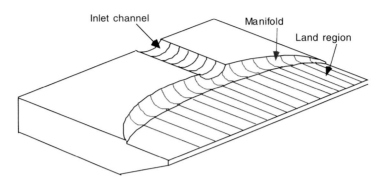

Figure 2.17 Coathanger manifold for sheet and flat film extrusion

- The die should be easy to assemble, disassemble, and clean.
- Spider supports should be located as far from the die exit as possible to minimize weld line problems.
- The land length is usually taken about ten times the die gap, although larger and smaller values can also work well.
- The total flow restriction along each streamline should be the same to achieve uniform exit velocities. An example is the coathanger die used in flat film and sheet extrusion; see Fig. 2.17. The length of the land reduces from the middle to the side to compensate for the longer flow path away from the middle. In cases where the flow restriction needs to be adjusted to fine-tune the extruded product, two methods can be used: mechanical and thermal. The mechanical adjustment involves a change in the flow channel geometry by a movable part, such as a choker bar, flex-lip, die plate, and so on. The thermal adjustment involves making a local change in the die temperature to increase or decrease the flow at that point. Some extrusion lines use automatic adjustment of flow restriction based on downstream measurement of the extruded product.

2.2.5 Devolatilization or Degassing

In some extrusion operations a vent opening in the barrel is used to remove volatiles from the plastic. Vented extruders require special extruder screws designed to create a zero pressure region under the vent port; this is necessary to keep the plastic melt from flowing out of the vent port — a condition called "vent flow." The screw that is used in a vented extruder is called a "two-stage" screw; see Fig. 2.18.

Vented extruders are often used with hygroscopic plastics to lower the moisture level to a level where it does not cause problems in the extruded product. Other volatiles can be removed too, such as monomers, solvents, air, and others. In many cases, removing volatiles through venting is more cost effective than a drying/separation operation before extrusion.

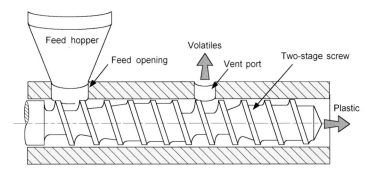

Figure 2.18 A two-stage extruder screw

A number of conditions have to be fulfilled to achieve efficient devolatilization:

- The plastic has to be completely melted by the time it reaches the extraction section under the vent port. Thus the vent port cannot be too close to the feed port.
- The pressure under the vent port has to be zero; therefore, the extraction section of the screw has to have large conveying capability and be only partially filled.
- The screw section just before the extraction section has to be fully filled with plastic melt. This is achieved by incorporating a screw section with low conveying capability.
- The screw section downstream of the extraction section (the second stage) has to have larger conveying capability than the screw section upstream of the extraction section (the first stage). This can be achieved by increasing the channel depth and/or pitch of the second stage relative to the first stage.

2.3 Efficient Extrusion

In the next two sections, we will discuss some of the critical aspects of machine design and process operation for efficient extrusion.

2.3.1 Efficient Machine Design

The components of an extruder were listed in Section 2.1. We will now list some of the most important aspects of these various components.

2.3.1.1 The Extruder Screw

The most important task of the extruder screw is to deliver a well-mixed plastic melt to the die at the required pressure and at a constant rate. The following is a listing of guidelines for good screw design:

- Use streamlined design, avoid dead spots.
- The screw should closely wipe the entire barrel surface to obtain good heat transfer, mixing, and a narrow distribution of residence times.
- The radial flight clearance between the screw and the barrel should be less than 0.003 times the screw diameter.
- A distributive mixing section should be incorporated to reduce consistency variations and melt temperature nonuniformities. The preferred location of the distributive mixing section is at the very end of the screw.
- A dispersive mixing element should be incorporated when the plastic contains solid filler particles that require high stresses to be broken down. Dispersive mixing elements can also be effective in making sure that unmolten plastic particles cannot travel to the end of the screw. Thus dispersive mixing elements can be useful even if there are no solid filler particles in the plastic.
- Mixing sections should have low pressure drop and preferably forward pumping capability.
- The screw, barrel, and die should be made from a corrosion resistant material when extruding PVC, fluoropolymers, or other plastics that expose the metal surfaces to corrosive attack.
- When the plastic contains abrasive fillers, such as titanium dioxide, glass, or others, the screw and barrel should be made out of a wear resistant material.
- When a screw coating is used, the preferred coating is one that has low friction properties. This will improve the conveying capability of the screw, resulting in higher output and better stability, and also the ease of cleaning the screw.
- For improved conveying and reduced hang-up of material, the flight flank radius should be large, while multiple flights and a small pitch should be avoided, see Fig. 2.19.
- Avoid abrupt changes in the channel depth along the length of the screw. An exception can be made in a multistage extruder screw where the transition sections just before and after the extraction section can be made quite short.
- When the extruder is equipped with a grooved barrel section, the screw must have a low compression ratio. The feed section should be shallow and the metering section relatively deep. Also, good mixing capability is critical in grooved barrel extruders.
- When a vent port is incorporated in the barrel, the length of the extruder should be increased to maintain reasonable melting and pressure generating capability.

- When the plastic melt viscosity is high or when the screw discharge pressure is low, the metering section can be made relatively deep. This also applies to situations where a gear pump is placed at the end of the extruder. Conversely, when the viscosity is low or the pressure is high, a shallow metering section should be used.
- When the bulk density of the feed material is low, the depth of the feed section should be made large (20% of the screw diameter or more). In some cases a crammer feeder may have to be used.

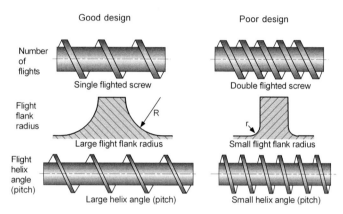

Figure 2.19 Methods of reducing friction on the screw and improve conveying

2.3.1.2 The Extruder Barrel

The extruder barrel has to contain the plastic and fit closely around the screw. The following characteristics are important for trouble-free operation:

- The barrel should be straight.
- The barrel should be designed to easily withstand the pressures that occur in the extruder. These can be as high as 70 to 140 MPa (10,000 to 20,000 psi) and even higher in injection molding.
- The barrel liner should be more wear resistant than the screw. The screw is easier to replace and less expensive to rebuild than the barrel.
- The downstream barrel support should be a sliding support, allowing the barrel to expand when heated. A rigid support can cause warping of the barrel with the possibility of severe damage to screw and barrel.
- For improved solids conveying, the feed section of the barrel can be grooved. Good cooling capability should be provided at the grooved barrel section to carry away high frictional heat and to prevent melt from accumulating in the grooves. Also, the grooves should taper gradually to zero depth to minimize the chance of material hang-up in the grooves.

- The vent port, if incorporated, should be offset and tangential to minimize the chance of plastic melt hanging up at the leading edge of the vent port opening, see Fig. 2.20. The more conventional symmetrical vent port design tends to give continuous problems with material building up at the bottom of the vent port. The vent port opening can be designed to slant downward to avoid having condensate enter the extruder barrel.

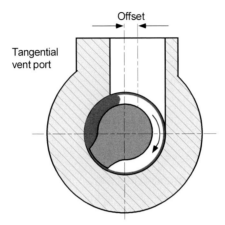

Figure 2.20 Good design of vent port

2.3.1.3 The Feed Hopper and Throat

The feed hopper should be designed to allow steady flow of the bulk material from the hopper to the extruder. The hopper should have smooth and gradual transitions and avoid sharp corners. The following characteristics are beneficial for stable flow:

- The side walls of the hopper should be steep to reduce the chance of bridging and piping; see Fig. 2.21.
- The cross-sectional shape should be circular, not square or rectangular; see Fig. 2.21; a circular geometry will minimize stagnation.
- A low friction coating on the inside hopper surface reduces hang-up of material.
- For difficult bulk materials, special features can be used to promote steady flow; examples are vibrating pads and crammer feed.

It should be noted that on many commercial extruders the geometry of the feed hopper and throat differs considerably from the desired geometry described above.

This partially results from the fact that sometimes machinery manufacturers put more emphasis on ease of manufacture than on functional performance. For processors, however, functional performance should be first priority in machine selection.

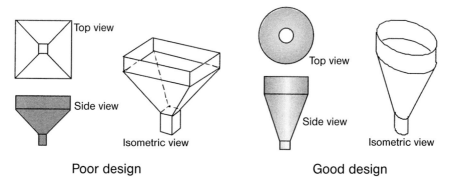

Figure 2.21 Poor and good design of feed hoppers

The feed throat should be designed to provide smooth flow from the feed hopper to the screw channel. The following characteristics are important:

- The feed throat should have good cooling capability to avoid premature heating of the plastic, which could cause the plastic to stick to the walls, resulting in lower output and unstable flow.
- At the barrel side of the feed throat should be a thermal barrier to minimize heat flow from the barrel to the feed throat housing.
- The axial length of the feed opening should be considerably larger than the screw pitch to avoid flight induced flow instabilities; see Fig. 2.22.
- The width of the feed opening should be smaller than the barrel diameter to maintain good forward flow; see Fig. 2.22.
- The feed opening should be offset, preferably tangentially, to maximize the intake capability of the screw; see Fig. 2.22.

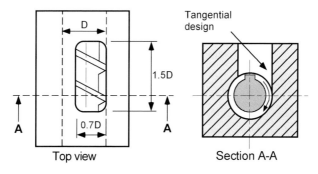

Figure 2.22 Good design of feed opening

2.3.1.4 The Extruder Drive

For the drive the following characteristics are important for efficient extrusion:

- The drive should have good speed regulation. Digital drives provide speed regulation of 0.01% full speed or better. Good speed regulation is particularly important when the extruder is operated at relatively low speed.
- The motor should have a direct coupling with the reducer. Many extruders use a belt transmission between the motor and reducer. However, these create a considerable power loss and cause slippage with resulting screw speed variation.
- An over-torque or over-current shutoff should be provided to prevent damage to the extruder.

2.3.1.5 Instrumentation and Control

The minimum requirements for instrumentation will be discussed in more detail in Chapter 6. Some important considerations are:

- The barrel temperature should be measured and controlled close to the inside barrel surface; thus a deep well temperature sensor should be used.
- Dual sensor temperature control can result in improved temperature control.
- On/off temperature control should be avoided in extrusion, whether used for heating or cooling.
- Melt temperature sensors should be in contact with the melt and designed to minimize conduction errors. Combined pressure/temperature transducers do not give a good indication of the melt temperature because the temperature sensor is not immersed in the melt stream.
- When PID (proportional/integral/derivative) control is used, the controller has to be carefully tuned under actual process conditions. Improper tuning can result in very poor temperature control. Some controllers set their tuning parameter automatically; these are called auto-tuning controllers.
- Capillary type pressure transducers filled with mercury cannot be used in extruded products used in food packaging or medical products.
- Piezo-resistive pressure transducers using silicon-on-sapphire technology can improve dynamic response, measurement accuracy and precision, and susceptibility to damage.
- A data acquisition system (DAS) is very useful in capturing data from the process and allowing careful analysis. Some systems can process the data and present statistical information such as control charts, standard deviation, process capability, alarms for presence of assignable causes, and so on. Such capability is very useful in implementing SPC and in process optimization [49].

2.3.2 Efficient Process Operation

An efficient extrusion process requires not only a good machine but also good operation of the process. Some important aspects of efficient extruder operation will be discussed next.

2.3.2.1 Feed Stock Consistency

It is impossible to have a stable process without having a consistent feed stock. At this point in time, the only way that a consistent feed stock quality can be ensured is by using material quality control (QC) to check critical material properties. Unfortunately, one cannot always rely on the resin supplier to deliver a material with consistent properties (see, e.g., [14]). There are a number of benefits to in-house testing. It allows the processor to establish a database on materials actually used in production. Thus nonconforming lots can be identified before problems occur on the production floor.

In-house testing allows a quantitative comparison of the material variability from different resin suppliers. This can provide useful selection criteria. Another benefit is that a company that practices materials QC is less likely to receive out of specification material (assuming the resin supplier knows that the material will be tested). Unfortunately, the opposite is true as well. Companies not practicing materials QC are more likely to receive out of specification material. Without materials QC it can be very difficult and costly to prove that a problem is caused by inconsistent feed stock properties.

One of the most commonly used material properties is the melt index (MI), also called melt flow index (MFI). Unfortunately, MI is one of the least useful properties because:

1. The measurement error can be quite large (± 20% or more), particularly for fractional MI plastics.
2. The MI is not a good indicator of the extrusion characteristics of a plastic.

A problem with the MI measurement is that only large variations in the resin can be positively identified. Smaller changes cannot because of the large measurement error. As a result, the MI measurement is not very discriminating. Unfortunately, the MI test is very popular with resin suppliers even though it is not very useful for processors. The melt index tester is shown in Fig. 3.1.

A better indicator of extrusion characteristics is the viscosity as a function of shear rate as determined on a rheometer, such as a capillary rheometer or cone-and-plate rheometer. Such data, however, are not always available from resin suppliers even though they are much more useful and discriminating than MI. It is possible to determine melt viscosity data directly from an extruder as discussed by Rauwendaal and Cantor [40].

A property that is very important in extrusion operations where the material is stretched, is the elongational viscosity. Examples of such extrusion operations are blow molding, fiber spinning, biaxially oriented film, blown film, and others. The elongational viscosity is usually not measured, even though simple tests are available to analyze the stretching behavior of plastic melts (see, e.g., [15]).

One of the important questions in materials QC is what properties to measure. Often properties are measured that may be important for end product performance but not for processibility; examples are environmental stress crack resistance, impact properties, flammability, among others. Properties that are important in extrusion are:

- Bulk density
- Compressibility
- Internal coefficient of friction
- External coefficient of friction
- Melting point
- Induction time at extrusion temperature
- Stabilizer level and type
- Viscosity versus strain rate and temperature
- Specific heat versus temperature
- Thermal conductivity/diffusivity
- Level of other additives (filler, lubricants, etc.)

Clearly, it is not always practical to measure all the properties important in extrusion. However, with a basic rheometer and a differential scanning calorimeter (DSC), many important properties can be readily determined. Bulk density and compressibility can be easily determined with a scale, a container, and a piston with weights. Thus good testing of processing related characteristics does not require an extraordinary investment in instruments or manpower. Most instruments, nowadays, are programmable with computer controlled data acquisition capability. This means that the manpower necessary to run these instruments is minimal.

2.3.2.2 Temperatures

Typical process temperatures for semicrystalline plastics are about 50 °C (122 °F) above the melting point. Amorphous plastics are usually processed about 100 °C (212 °F) above the glass transition temperature. On the one hand, we would like to keep the process temperatures high to reduce the melt viscosity and make the material more easily processible. On the other hand, we would like to keep the temperatures low to reduce the chance of degradation. As a result, a thermally stable plastic may have a wide operating window. However, a thermally sensitive plastic may have only a narrow temperature range within which it can be successfully processed.

The temperature profile along the extruder can have a strong effect on the extruder performance. The following guidelines can be used to set the temperature profile:

- Set the barrel temperature in the feed section such that the diehead pressure fluctuation is minimum. This requires a graphical display of the pressure as a function of time, either using a chart recorder or a CRT (cathode ray tube).
- Set the transition section to the temperature where the melt temperature variation is minimum.
- Set the metering section and die temperature at the value at which the melt temperature should be controlled.

Some extruders have a melt temperature feedback control. This typically changes the set point of the last one or two barrel temperature zones in response to a change in melt temperature. Such a system can only react to very slow changes in melt temperature. It should be noted also that reduced barrel temperature in the metering section will increase the pumping capacity of the metering section. Also, increased screw temperature increases the metering section pumping capacity.

2.3.2.3 Screen Pack

In many extrusion operations a screen pack is used to filter contaminants so that they do not end up in the final product. A drawback of the screen pack is that its flow restriction increases as contamination builds up. This will cause a reduction in output. Unless the take-up is reduced correspondingly, this will change the product dimensions. One way to approach this is to use a pressure feedback control that changes the screw speed in such a way as to keep the pressure behind (downstream) the screen pack constant; see Fig. 2.23.

Another approach is to change the screen pack before the pressure drop across the screens changes too much. This may not work too well because it can lead to frequent screen changes (increased downtime and scrap). Also, small changes in pressure can cause relatively large changes in output.

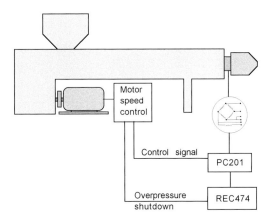

Figure 2.23 Pressure feedback control

A better approach would be to use an automatic screen changer that allows a rapid screen change, preferably without having to stop the line. Yet another possibility is to use a gear pump to control the output. With a gear pump the screw speed is usually varied in such a way as to maintain constant inlet pressure to the gear pump.

2.3.2.4 Feeding

Most single screw extruders are flood fed. Thus the screw takes in as much material as it can. In this case, the screw is essentially completely filled with plastic and the output is determined by the screw speed. Another approach is to use starve feeding; see Fig. 2.24.

In starve feeding the feed material is metered into the feed throat at a rate below the flood feed rate. The output of the extruder is no longer determined by the screw speed but by the feed rate set for the feeder. In starve feeding the material does not build up in the feed hopper. As a result, this is one way to avoid flow problems in the feed hopper.

The extruder is only partially filled in starve feeding; the degree of fill is dependent on the feed rate and the screw speed. Increasing the screw speed reduces the degree of fill, while raising the output will increase the degree of fill. By changing feed rate or screw speed, the effective *L/D* (length to diameter ratio) of the extruder can be varied. A number of advantages can be obtained from starve feeding:

- The motor load and screw torque can be reduced.
- The specific energy consumption (SEC) can be varied.
- Instabilities may be eliminated, for example, solid bed breakup.

The SEC determines how much mixing and viscous heat generation occurs in the plastic. Starve feeding, thus, allows an additional degree of process control. It can be very useful in cases where excessive motor load and/or excessive melt temperatures occur.

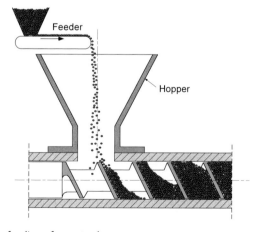

Figure 2.24 Starve feeding of an extruder

Another feeding technique that can be quite useful is gravimetric output control (GOC). In GOC, small amounts of feed material are weighed and discharged with a measurement of the time elapsed between each discharge. This is typically about 15 to 30 seconds. Thus the mass flow rate is measured about every 15 to 30 seconds. If the mass flow rate changes, either the screw or the line speed can be adjusted to maintain the same mass per length of extruded product. GOC has found widespread use, particularly in multilayer coextrusion where it is used to control the average thickness of each individual layer. GOC usually allows an output control of 0.5% which is quite good. In a way, GOC can be seen as an alternative to a gear pump, although it is quite a different system. GOC can be used in certain applications where a gear pump might present a serious problem, for example, when the plastic contains a high level of abrasive filler or when the plastic has limited thermal stability.

2.3.2.5 Gear Pumps

Gear pumps consist of two closely intermeshing gears that fit closely in the housing; see Fig. 2.25. The gear pump is, in essence, a counterrotating twin screw extruder with very short screws and very large flight pitch.

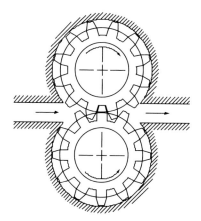

Figure 2.25 Gear pump

Gear pumps offer a number of benefits:

- Good output stability, output variation less than 1.0%
- High volumetric efficiency, usually better than 95%
- Good pressure generating capability
- Agglomeration of filler can be reduced

On the other hand, there are some disadvantages as well:

- Poor mixing capability
- Low energy efficiency, typically 20 to 50%
- Susceptibility to wear
- Expense

Good applications for gear pumps are

- To generate pressure for extruders that have low pressure generating capability, such as nonintermeshing twin screw extruders, corotating twin screw extruders, or vented single screw extruders.
- To maintain good output uniformity in processes where the product dimensions have to be tightly controlled. Examples are fiber spinning and medical tubing. A regular extruder usually can keep output fluctuations as low as 1%. A gear pump can keep output fluctuations to less than 1%, from about 0.2 to 0.5% in the best case.

In the following situations there is a substantial chance of problems when a gear pump is used:

- When the plastic contains a high level of abrasive filler
- When the plastic has poor thermal stability

The reason that degradation can easily occur in gear pumps is that not all the plastic melt flows in the spaces between the teeth where it is conveyed efficiently. A small amount of melt flows to the sides of the gears where it is used to lubricate the gears. This lubrication flow can cause a small amount of plastic to remain in the gear pump for a long time. The long residence time combined with the high temperature in the gear pump will make degradation likely to occur.

3 Plastics and Plastics Properties Important in Injection Molding and Extrusion

3.1 Thermoplastics and Thermosets

Plastics can be divided into thermoplastics and thermosets. Plastics are also called polymers; they are materials made up of very large molecules. Thermoplastics are materials that soften or melt and flow as a thick fluid when heated above a certain temperature. In this state, the material is often referred to as a plastic melt. It is also in this state that the material is usually formed or shaped into a product. Upon cooling thermoplastics harden and behave as a solid. After a thermoplastic product has been formed, it can be reheated and softened to be shaped again. Thus thermoplastics can be processed several times, and this is what makes them suitable for recycling.

Thermosets are materials that harden when heated above a certain temperature. The hardening is due to a curing or crosslinking reaction that connects the individual molecules and causes the formation of a three-dimensional molecular network. The shaping of thermosets usually occurs before the crosslinking sets in, thus at a temperature below the curing temperature. The crosslinking reaction is not reversible; a thermoset cannot be softened again like a thermoplastic. It is more difficult, therefore, to recycle a thermosetting material than a thermoplastic. Examples of thermosets are phenolics, ureas, certain polyesters, melamines, and alkyds.

3.2 Amorphous and Semicrystalline Plastics

Thermoplastics can be further divided into amorphous and semicrystalline plastics. Amorphous plastics have a random, irregular molecular structure. Examples of amorphous plastics are polystyrene (PS), polycarbonate (PC), acrylic (PMMA), acrylonitirile butadiene styrene (ABS), and polyvinylchloride (PVC). Semicrystalline plastics can form highly regular regions where the molecules form crystals; these crystalline regions are referred to as crystallites. The ability to form crystals is determined to a large extent by the shape of the plastic molecule. Plastics that have linear molecules without large sidegroups usually have the ability to form crystallites. An example is high density polyethylene (HDPE), which can achieve levels of crystallinity as high as 90%. Other polymers that can form crystalline regions are acetal (POM), nylon (PA), polyester terephthalate (PETP), low density polyethylene (PE-LD), and polypropylene (PP). Plastics with bulky sidegroups often cannot form crystallites and, therefore, are amorphous. An example is polystyrene. Acronyms for common plastics are given in Appendix I.

The crystalline regions in thermoplastics have different properties than the amorphous regions; for instance, the density and the optical properties are different. As a result, the light transmission through a plastic changes when crystallites are present; the crystallites act as a filler and make the material opaque or translucent below the melting point. Above the melting point, the crystallites disappear and the material is transparent. Since amorphous plastics have no crystallites, they are often transparent — unless of course they contain fillers or other materials that alter their optical properties. It is interesting to note that "crystal polystyrene" is an amorphous plastic. It is called "crystal" because it is transparent, not because it is crystalline.

Semicrystalline plastics are never completely crystalline; the highest level of crystallinity occurs in high density polyethylene. Despite this, semicrystalline plastics are often referred to as crystalline material. It should be remembered, however, that the term "semicrystalline" is more appropriate. Some plastics crystallize rapidly, such as high density polyethylene, while others crystallize slowly, for example, polyethylene terephthalate (PET). In fact, if PET is quenched rapidly after melt forming, it may cool down in a completely amorphous state. In general, the morphology that develops in a plastic will depend on how fast it is cooled during and after the shaping process. The morphology is also affected by the stresses exerted on the plastic during and after the shaping process. Thus the flow and temperatures in the tooling play an important role in the morphology that ultimately develops in the plastic part. The part properties are strongly determined by the morphology of the part, and thus the properties are affected by the flow and temperatures in the tooling.

3.3   Liquid Crystalline Plastics

Liquid crystalline plastics (LCPs) are a special class of plastics. The molecules of LCPs are rod-like structures organized in large parallel domains; this is true not only in the solid state but also in the melt state. The large, ordered domains give LCPs unique characteristics compared to amorphous and semicrystalline plastics.

Differences in mechanical and physical properties between plastics can often be attributed to their structure. The order in semicrystalline plastics and LCPs make them stiffer, stronger, and less resistant to impact than amorphous plastics. Semicrystalline and liquid crystalline plastics tend to be more resistant to creep, heat, and chemicals; however, they tend to require higher melt temperatures in processing and shrink more during cooling than amorphous plastics. As a result, they are more susceptible to warpage in injection molding.

When amorphous plastics are heated, they soften gradually, while semicrystalline plastics tend to soften more abruptly. Amorphous plastics tend to not flow as easily as melted crystalline plastics in melt processing. LCPs have the high melt temperature of semicrystalline plastics, but they soften gradually like amorphous plastics. LCPs have the lowest viscosity, shrinkage, and warpage of all thermoplastics. Some of the general characteristics are summarized in Table 3.1.

Table 3.1 General Characteristics of Various Plastics

Property	Amorphous	Semicrystalline	Liquid crystalline
Density	Lower	Higher	Higher
Tensile strength	Lower	Higher	Highest
Tensile modulus	Lower	Higher	Highest
Ductility	Higher	Lower	Lowest
Creep resistance	Lower	Higher	High
Maximum use temperature	Lower	Higher	High
Shrinkage/warpage	Lower	Higher	Lowest
Flow	Lower	Higher	Highest
Chemical resistance	Lower	Higher	Highest

3.4 Elastomers

Thermoplastic elastomers generally are low modulus materials. They can be stretched to at least twice their original length at room temperature and return to their approximate original length when the deforming force is removed. Thermoset rubber materials have been around for a long time, but nowadays many injectionmoldable thermoplastic elastomers (TPEs) are replacing traditional rubbers. TPEs are also used to modify the properties of rigid thermoplastics, in most cases to improve their impact strength.

3.5 Flow Behavior of Plastics

In order to understand how a plastic behaves in the injection molding and extrusion process, it is helpful to know how the plastic flows in the melt state. Since plastics are made up of very long molecules, they do not flow very easily in their melt state. The ease with which a fluid flows is described by the viscosity of the fluid. Water flows quite easily, but honey flows less easily; the viscosity of honey is larger than that of water. The viscosity is often expressed in the units poise; the unit pascal·second is also used. It is easy to convert from poise to pa·s: 10 poise = 1 pa·s. To get an idea how the viscosity of plastic melt compares to other materials, Table 3.2 shows the approximate viscosity of various materials, expressed in pa·s.

Table 3.2 The Viscosity of Various Materials

Material	Viscosity (pa·s)
Air	0.00001
Water	0.001
Olive oil	0.1
Plastics melts	100 to 1,000,000
Pitch	1,000,000,000

It is clear from Table 3.2 that the viscosity of plastics is much higher than the viscosity of water, by at least five orders of magnitude! With a higher plastic viscosity, more torque is required on the extruder screw of the plasticating unit and more pressure is necessary to force the plastic melt into the mold. The viscosity of a plastic is very much dependent on the molecular weight of the plastic; the higher the molecular weight, the higher the viscosity. Since for one plastic, for example polyethylene (PE), there are many grades with different molecular weights, the viscosities can vary substantially from one PE to another.

3.5.1 The Melt Index Test

The flowability of a plastic is often measured in a melt index tester. The melt index machine is a simple ram extruder; see Fig. 3.1.

Plastic is placed in the reservoir and heated to the appropriate temperature. A weight is placed on top of the ram and this causes the plastic melt to be extruded out of the melt index die located at the bottom of the reservoir. The melt index (MI), sometimes called the melt flow index (MFI), is the amount of plastic extruded in grams in a certain time period, usually 10 minutes.

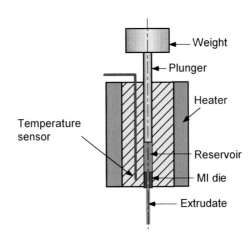

Figure 3.1 A schematic of the melt index apparatus

In the melt index test, a low viscosity plastic will flow out faster than a high viscosity plastic. Thus a high MI is associated with a low viscosity plastic and a low MI with a high viscosity plastic. The term fractional melt plastic is often used; this means that the plastic has a melt index less than one. A melt index less than one is considered low, and thus fractional melt plastics have high viscosity.

3.5.2 Spiral Length Tester

A flow test that is used specifically in the injection molding industry is the spiral length test. In this test a plastic melt in injected into a mold with a spiral flow channel; see Fig. 3.2.

The length that the plastic can travel along the spiral channel is determined by the viscosity of the plastic. The lower the viscosity, the longer the spiral fill length will be. Obviously, the injection conditions will have to be standardized for the spiral length data to be comparable.

Figure 3.2 A spiral mold test mold

3.5.3 The Effect of Shearing

When a plastic is processed in an extruder or in a mold, it is sheared in the flow that occurs. This is due to different layers of the plastic moving at different velocities. The rate of shearing that occurs in a fluid is called the shear rate; it is the difference in velocity between two fluid elements divided by the distance between the elements; see Fig. 3.3.

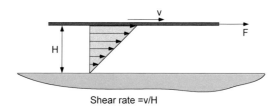

Figure 3.3 Shearing of a fluid between two parallel plates

The shear rate is determined by the flow rate and the geometry of the flow channel. When the flow rate of the plastic is high, the shear rates will be high. Also, when the flow channel is small, the shear rate will be high.

3.5.3.1 Shear Thinning or Pseudoplastic Behavior

In plastics, the viscosity changes when the shear rate changes; a fluid that behaves that way is called a non-Newtonian or nonlinear fluid. This behavior is due to the fact that the plastic molecules are very long and entangled. The entanglements of the molecules determine the viscosity of a plastic. When a plastic is exposed to a high shear rate, the number of entanglements of the molecules reduces and with it the viscosity. When the shear rate reduces, the viscosity increases again. This behavior is called *shear thinning behavior*; it is also called *pseudoplastic behavior*. This behavior is very important in extrusion and injection molding.

The effect of shear rate on viscosity can be very strong. In certain plastics, the viscosity can reduce several orders of magnitude when the shear rate is raised. This is particularly important in injection molding where local shear rates can be very high, typically in the gate region where the channel is very small and the flow rate quite high. In a highly shear thinning plastic, the viscosity will reduce a great deal in this region and this will keep the injection pressure relatively low. Thus a high degree of shear thinning is beneficial in injection molding.

The relationship between viscosity and shear rate is often described by a power law equation, see Fig. 3.4.

A fluid that follows such a relationship is called a power law fluid. The power law equation has two important parameters: the consistency index and the power law index. The consistency index is value of the viscosity at a shear rate of one. The power law index is a measure of the degree of shear thinning behavior; for plastics it varies between zero and one. The closer the power law index is to zero, the more strongly shear thinning the plastic. When the power law index is close to one, the plastic is only slightly shear thinning. When

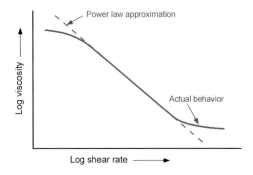

Figure 3.4 Viscosity versus shear rate, actual behavior and power law approximation

the power law index equals one, the viscosity is not affected by shear rate; a fluid that behaves this way is called Newtonian.

3.5.3.2 Effect of Temperature on Viscosity

When the temperature of a plastic melt is increased, the viscosity reduces. This is illustrated in Fig. 3.5.

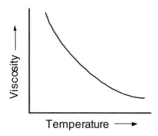

Figure 3.5 Effect of temperature on melt viscosity

The effect of temperature on viscosity varies from one plastic to another. In general, amorphous plastics have a high temperature sensitivity relative to semicrystalline plastics. The temperature coefficient for amorphous plastics ranges from about 5 to 20%. This means that the viscosity changes from 5 to 20% for each degree centigrade change in temperature.

For semicrystalline plastics the temperature coefficient of the viscosity is about 2 to 3%. A change in the extruder barrel or mold temperature is going to have a larger effect on an amorphous plastic than on a semicrystalline plastic. Good temperature control, therefore, is even more critical in amorphous plastics than in semicrystalline plastics.

3.5.3.3 Effect of Pressure on Viscosity

The effect of pressure on viscosity is small at moderate pressures. In injection molding, however, very high pressures can occur, as high as 140 to 210 MPa (20,000 to 30,000 psi) and higher. At those pressures the viscosity can change considerably due to pressure; see Fig. 3.6.

3.5.4 Flow Properties for Injection Molding

In injection molding the injection pressures are often quite high due to the fact that the plastic melt has to be forced through small flow channels; this is particularly true at the gates. In injection molding the viscosity of the plastic at normal process temperatures and high shear rates should be low enough to avoid excessive pressures. The plastic may have a

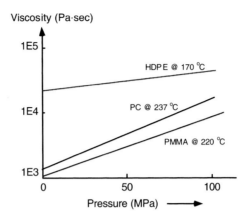

Figure 3.6 The effect of pressure on melt viscosity for three plastics

high viscosity at low shear rate, indicated by a high consistency index; however, the viscosity at high shear rate should be low, indicated by a low power law index.

3.5.5   Viscous Heat Generation

When the plastic melt is sheared, heat is being generated in the plastic. This is called viscous heat generation. Viscous heat generation is determined by the product of viscosity and shear rate squared. Thus, the higher the viscosity of the plastic, the higher the viscous heat generation. The same is true for the shear rate; however, the shear rate has a stronger effect since the viscous heating increases with the shear rate squared.

As a result of the high viscosity of plastics, most of the heating of the plastic in the plasticating unit typically comes from viscous heat generation. In fact, in some cases too much viscous heat generation occurs in the reciprocating extruder and the machine has to be cooled to maintain the desired melt temperatures.

3.6   Thermal Properties

Understanding the thermal properties of a plastic is important in understanding how a particular material behaves in an injection molding machine or extruder.

Knowledge of the thermal properties allows the selection of the appropriate machine and setting of the correct process conditions; it also helps in analyzing process problems.

3.6.1 Thermal Conductivity

Several thermal properties are important in injection molding. One of the most important properties is thermal conductivity. This is the ability of a material to conduct heat. Plastics have a low thermal conductivity — they are considered to be thermal insulators. This means that heating and cooling plastics by conduction is a slow process. Heating occurs in the plasticating unit, usually a reciprocating extruder, and cooling occurs in the mold; thermosets are an exception; they are heated in the mold. The low thermal conductivity often determines how fast a plastic can be processed. This is true not only in injection molding but also in extrusion and, in fact, in most plastic processing operations.

3.6.2 Specific Heat and Enthalpy

The amount of heat necessary to increase the temperature of a material by one degree is given by the specific heat. In most cases, the specific heat of semicrystalline plastics is higher than that of amorphous plastics. The amount of heat necessary to raise the temperature of a material from a base temperature to a higher temperature is determined by the enthalpy difference between the two temperatures. If we use room temperature as the base temperature, the enthalpy of different plastics can be plotted against temperature; this is shown in Fig. 3.7.

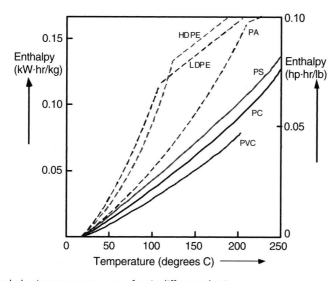

Figure 3.7 Enthalpy/temperature curves for six different plastics

Enthalpy is expressed in kW·hr/kg or hp·hr/lb; it is a specific energy, in other words, energy per unit mass. Most of the energy required in the processing of plastics is needed to increase the temperature of the plastic. If we know the starting temperature, usually room temperature, and the discharge temperature, we can determine the minimum energy required to process the plastic. For instance, if we look at PVC in Fig. 3.7, then we can see that the specific energy required to raise the temperature from room temperature to 150 °C is about 0.03 hp·hr/lb. Thus for each lb/hr, we require 0.03 hp. If we process PVC starting at room temperature up to 150 °C at 100 lb/hr, the minimum power requirement is 3 hp.

If we compare low density polyethylene (PE-LD) to PVC, we see that PE-LD requires about 0.09 hp·hr/lb to go from room temperature to 150 °C. Thus the specific energy requirement for PE-LD is much higher than that for PVC. In general, semicrystalline plastics have higher specific energy requirements than amorphous plastics. Obviously, this affects the cooling as well. It means that to cool PE-LD from 150 °C to room temperature much more heat has to be removed than in cooling the same mass of PVC at 150 °C down to room temperature.

3.6.3 Thermal Stability and Induction Time

Plastics can degrade in the injection molding process. The main variables involved in degradation are temperature and the length of time that a plastic is subjected to an elevated temperature. Plastics degrade when exposed to high temperatures; when the temperature increases, the degradation occurs more rapidly. Degradation can result in loss of mechanical properties, optical properties, appearance problems, degassing, burning, and so on. Other variables can affect degradation, for instance the presence of oxygen.

A quantitative measure of the thermal stability of a plastic is the induction time. This is the time at elevated temperature that the plastic can survive without measurable degradation. The longer the induction time at a certain temperature, the better the thermal stability of the plastic. The induction time can be measured using various instruments, such as a TGA (thermo-gravimetric analyzer), TMA (thermomechanical analyzer), cone-and-plate rheometer, or others.

When the induction time is measured at several temperatures, the induction time can be plotted against temperature. Such a graph is shown in Fig. 3.8.

3.6.4 Density

The density of a material is the mass divided by the volume of a sample of the material; it is often expressed in gram per cubic centimeter or gr/cc. The density of most plastics is about the same as the density of water, which is 1 gr/cc. The densities of various plastics at room temperature are shown in Table 3.3.

3.6 Thermal Properties

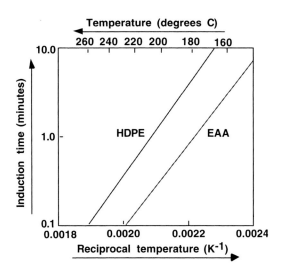

Figure 3.8 Induction time of HDPE and EAA as a function of temperature

Table 3.3 Density Values for Various Plastics

Plastic	Density in (gr/cc)
LDPE	0.92
HDPE	0.95
PVC	1.40
ABS	1.02
PP	0.91
NYLON-6	1.13
PETP	1.35
PS	1.06
FEP	2.15

The density can also be described by the term specific volume; this is the reciprocal of the density. Thus it is volume divided by mass and can be expressed in the units cc/gr. The density and the specific volume are affected by temperature and pressure. The mobility of the plastic molecules increases with higher temperatures. As a result, the specific volume increases with increasing temperature, this is illustrated in Fig. 3.9 for HDPE. The specific volume reduces with higher pressures; this shown in Fig. 3.9 as well.

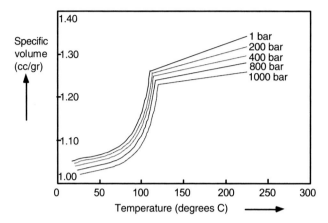

Figure 3.9 Specific volume of HDPE versus temperature at various pressures

The volume of a certain mass of plastic changes with both temperature and pressure. During the injection molding process both the temperature and the pressure vary substantially. The effect of both pressure and temperature on the volume of a plastic is often evaluated by using the PVT diagram; Fig. 3.9 is an example of a PVT diagram.

4 Introduction to Statistical Process Control

4.1 Introduction

What is statistical process control or SPC?

SPC simply means the use of statistical methods to monitor, analyze, and control a process, in particular process variations.

SPC can be applied to many different processes. For instance, to manufacturing processes, such as the production of electronic chips, batteries, tires, plastic bottles, and so on. However, it can also be applied to tasks performed in the service industries, such as filling out claim forms, diagnosis of patients by a medical doctor, processing of tax refunds by the IRS, processing of mail by the Post Office, and so on. A term closely related to SPC is statistical quality control (SQC). Statistical quality control is the application of statistical methods to measure and improve the quality of the process and the articles produced.

Statistics deals with the collection, organization, analysis, and interpretation of numerical data. An integral part of statistics is mathematics; this can make statistics rather imposing and unappetizing. However, the basic mathematical tools necessary to do SPC are straightforward and do not need to be intimidating at all. In the following sections, mathematics will be kept to a minimum to make the material accessible to as many people as possible. Appendix IV will cover more mathematical detail for those so inclined.

The goals of SPC are to improve and ensure quality, and thus reduce process cost due to waste as a result of rejects. Quality is determined by how much the characteristic of a product, for instance the outside diameter of a pipe, differs from the target value. The closer the product characteristic is to the target value, the higher the quality.

The aim of process control is to prevent defects. On the other hand, the more traditional product control is aimed at detection of defects. There are subtle but very important differences between process and product control; see Table 4.1.

Table 4.1 Product Control Compared to Process Control

	Product control	Process control
Focus	Product	Process
Goal	Variability within specification limits	On target with smallest variation
Tools	Acceptance sampling	Control charts
Improvement	Outgoing quality only	Quality plus productivity
Philosophy	Detection and containment of problems	Prevention of problems

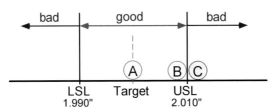

Figure 4.1 Illustration of detection based quality control

The goal of product control is to make sure that the variability of a product characteristic is within the specification limits. This means that the product characteristic has to be below the upper specification limit (USL) and above the lower specification limit (LSL); see Fig. 4.1.

The product characteristic in Fig. 4.1 is a length with USL = 2.010 in, LSL = 1.990 in, and a nominal target value of 2.000 in. Product A with a length of 2.000 in will of course be accepted. Product B with a length of 2.009 in will also be accepted; however, product C with a length of 2.011 in will not be accepted. In a detection-based quality control system, products A and B would both be considered good, while product C would be considered bad. It is clear, however, that products B and C are much closer together than products A and B. In fact, product A is much better than product B because A is right at the target value, while B is far from the target value, actually very close to the USL. In a prevention-based quality control system, product A is considered best, product B fair, and product C bad; see Fig. 4.2.

The main tool in product control is acceptance sampling by inspection of finished goods. Some problems with this traditional approach to quality control are:

- Inspection plans usually do not catch all the rejects.
- When items are rejected the damage is already done.
- Inspection is time consuming and costly.

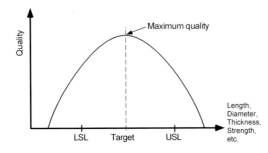

Figure 4.2 Quality in a prevention based quality control system

A rejected product may be reworked to make it acceptable. However, there is additional cost associated with this rework. In the worst case, the product cannot be reworked and becomes an expensive piece of scrap.

Statistical process control involves quality control in each step of the production process. Standards are established for each step and an acceptable range for each standard is determined. As long as each process step yields a product within its set range, quality is ensured. When the acceptable range is exceeded, or a trend is identified that indicates the range will soon be exceeded, the process step is stopped and adjusted to bring it back in line with the standard.

It should be realized that SPC cannot solve all quality and production problems. It is not a cure-all! It will not correct a poor product design or poor employee training. Nor will it correct an inefficient process or worn-out machines and tooling. However, it can help detect all of these types of problems and identify what corrective action is necessary to solve the problem.

One of the most important tools in SPC is the control chart. The most common control chart shows how the average of a set of measurements and the range of the measurements vary with time; see Fig. 4.3. The *average* of a set of measurements is often called the *mean* and represented by the symbol $\bar{x}$ pronounced x bar and sometimes written as x-bar. The *range* is the difference between the largest and smallest measurement in a group of data; it is represented by the symbol R. The chart shown in Fig. 4.3 is called an $\bar{x}$ and R chart or the Shewhart control chart. Control charts can be used for controlling processes, for identifying problems, and for spotting trends before they become problems. Control charts will be covered in more detail in Chapter 7.

The name of *Dr. Walter A. Shewhart,* mentioned above, is strongly linked to SPC. In fact, Dr. Shewhart can be considered the father of SPC. In the 1920s, Dr. Shewhart developed methods for statistical monitoring of manufacturing processes. These methods, with very few changes, are still commonly used today. Dr. Shewhart's book *Economic Control of*

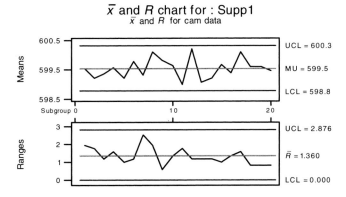

Figure 4.3 Example of an $\bar{x}$ and R chart

Quality of Product [1], published in 1931, was the first book to describe the basic elements of statistical process control.

Another name often linked to SPC is that of *Dr. W. Edward Deming*. Dr. Deming, a former colleague of Dr. Shewhart, was instrumental in introducing SPC in Japan after the Second World War. Ironically, even though SPC was developed in the US, it was largely ignored in the US, while Japan embraced the concepts of SPC under the tutelage of Dr. Deming. It is clear that one of the reasons for Japan's industrial success is the widespread acceptance and use of statistical process control.

As a result of increased competitive pressure, SPC has made a resurgence in the US in the 1980s. Some companies embrace SPC because they realize that it is a powerful tool to improve quality and productivity. Other companies adopt SPC because their customers require suppliers to use SPC and to supply SPC data with the product to prove that their production processes are in control and capable of making good products consistently. An example is the American automobile industry; its manufacturers have not only implemented SPC in their own production facilities but also demanded that their suppliers implement SPC methods to ensure quality of subcontracted automobile components.

Another important name in SPC is that of *Dr. Dorian Shainin*. Shainin was part of a team that developed the concepts of precontrol, a method of SPC much simpler than the classical SPC method. Shainin also developed a whole set of tools for use in design of experiments. Although Shainin is not as well known as Shewhart and Deming, his contributions have been no less important. Several major US corporations have used his techniques to achieve world-class quality. A good description of the Shainin techniques is given in Bhote's book *World Class Quality, Using Design of Experiments to Make It Happen* [19].

4.2   Implementing Statistical Process Control

One of the key issues in implementing SPC is training. Training has to occur at all levels: production workers, supervisors, and management. SPC gives production workers the tools to monitor their efforts and to improve the product being made, enhancing both pride in workmanship and job satisfaction.

Supervisors need to learn SPC so that they can make sure it is properly implemented on the work floor and so they can interpret SPC data in troubleshooting process problems. While SPC gives workers the proof that their work meets quality standards, it also gives both workers and supervisors the key to unlock the source of quality problems.

Management needs to learn SPC so that they can interpret SPC information in monitoring the output of their business. An added benefit is that statistical quality data can be a valuable product-marketing tool. The SPC data gives management the proof of the quality of their product. As mentioned earlier, many companies require proof of quality from their suppliers of everything from raw material to semifinished and finished goods. A properly documented SPC program provides proof of quality required to obtain those contracts.

The bottom line of SPC is profits. More profits arise from improved quality and productivity. Workers are given the tools to work smarter, not harder, to ensure quality with the goal of no rejects and no wasted effort. Supervision is given the means to coordinate production and quality control while eliminating the traditional antagonism between production and quality control functions. Management is given the tools to manage quality and reduce waste. The biggest selling point of SPC is more quality output for the same amount of input — that is, increased productivity. Various tools are available to implement a statistical process control program. These tools include problem identification techniques, process control charts, and capability measures. Applying these tools, however, is no guarantee that the benefits of SPC will be attained. These tools are all used to find problems. The benefit of an SPC program is increasing quality and productivity through *finding and eliminating* problems.

One of the primary causes of failure of SPC is giving workers the tools to find problems without the means to solve these problems. Not being able to solve problems leads to inefficiency and frustration. Dr. Deming suggests that when a worker identifies a problem on the production floor, the person immediately responsible for that process is able to take corrective action and solve the problem only 6% of the time. The other 94% of the time the problem must be solved through *cooperative action* of various individuals: operators, supervisors, engineers, managers, and so on. This means that effective management of SPC efforts is key to the successful implementation of SPC. The process of implementing SPC requires several steps to be taken.

The first step is the creation of an environment that allows and encourages problem solving. This step is often the most difficult one. SPC allows accurate detection of problems. However, detection of a problem is only part of the process of problem elimination. The next part of the process is recognition and acknowledgment of the problem by the appropriate people. This invariably includes a person with supervisory responsibilities. In some cases, a supervisor or department manager may be reluctant to acknowledge a problem. This may be because he or she is afraid this creates a bad impression outside the department. The manager may then respond, not by acknowledging the problem, but by turning against the person who reported the problem. By turning against the bearer of bad news, the department manager has most likely made sure that nobody in the department will report a problem again. Consequently, this manager will be largely ignorant about problems in the department, making effective management next to impossible.

Creation of an atmosphere that allows problem solving, therefore, requires enlightened and capable supervisors and managers. This is an absolute necessity for efficient implementation of SPC. People should be encouraged to uncover and report problems. People reporting problems should be praised and rewarded; they should not be treated as troublemakers. The real troublemakers are the people ignoring problems, not the people reporting them!

The second step is training in SPC and process technology. In order to apply SPC successfully, the people involved with the process should have a basic familiarity with SPC concepts. As discussed earlier, training has to occur at all levels. However, training should cover not only statistical process control but also process technology. It is very difficult to apply SPC to injection molding if there is not a good understanding of the injection molding process. The same is true for other processes, such as extrusion, blow molding, compounding, and so on. Therefore, this book covers not only basic SPC concepts but also basic injection molding technology and extrusion technology.

The third step is to determine the key process problems and variables. Various methods can be used to identify and prioritize problems in SPC; these include: brainstorming sessions, fishbone (cause and effect) diagrams, Pareto charts, histograms, scatter diagrams, and so on. These methods will be discussed in detail in the next chapters. It is important to direct efforts to implement SPC to the most significant problems where an improvement in quality and productivity can be made and measured.

The fourth step is to make sure the measurement system is capable of measuring the level of variation that is occurring in the process. In any measurement system there is some level of measurement error. It is important to ascertain that the measurement error is small relative to the variation that we are trying to measure. This will be discussed in Chapter 6 on measurement. A lot of information can be gathered and many control charts can be generated, but if the measurement system is not capable, all of this will be essentially useless.

The fifth step is to use SPC to bring the process into control with its preset set of conditions. These conditions can typically be grouped under the following main categories (the six m-words):

- Man (personnel)
- Machine
- Material
- Methods
- Measurement
- Milieu (environment)

In a Chapter 7 it will be shown how process control charts can be used to reduce and eliminate causes of variation. The most commonly used control chart is the $\bar{x}$ and R chart shown earlier in Fig. 4.3.

The sixth step is to determine process capability. Once the process has been brought into control, it has to be determined whether the characteristics of the individual products meet the requirements of each characteristic. These requirements are often stated as engineering limits or product specifications. An example is a molded cover with a nominal thickness of 0.020 in with an allowable upper limit of 0.021 in and lower limit of 0.019 in. In this case the upper specification limit, USL, is 0.021 in and the lower specification limit, LSL, is 0.019 in. Specification is often abbreviated as spec.

If the actual product variability is less than the specification width, the process to make the product is considered *capable,* see Fig. 4.4.

If the product variability is larger than the specification width, the process is considered not capable, see Fig. 4.5.

In this case nonconforming products will be produced, which will result in rework or rejects. Various measures of process capability will be discussed in Chapter 8.

If the process is not capable, some corrective action is necessary. Thus, step seven is to implement a plan of process improvement. Various approaches are possible:

- Improve process to reduce product variability.
- If the specification width is unnecessarily tight, the specification width could be broadened.
- Measure each product and eliminate out-of-specification products.
- Let the customer separate bad from good products.
- Quit making the product.

The last three approaches are usually not desirable. Broadening the specification width is often not possible. The most desirable corrective action, therefore, most often is to improve the process. This will typically involve a change in one of the conditions affecting the

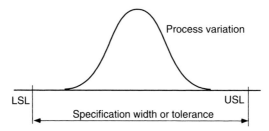

Figure 4.4 Example of a capable process

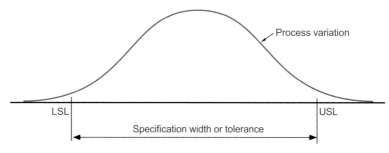

Figure 4.5 Example of a process that is not capable

process. These are, as discussed earlier: man, machine, material, methods, measurement, and milieu; see Fig. 4.6.

After a change has been made, we move back to step three and repeat the process until the process becomes *capable;* see Fig. 4.7. In other words, repeat the process until the product meets requirements consistently.

Next, we have to determine whether the products meet requirements for all important product characteristics; dimensions, color, surface finish, impact strength, tensile strength, and so on. If the requirements for one characteristic are not met, then we have to take the next step. The eighth step is to identify the next important problem. We then move back to step two and repeat the various steps until all requirements can be met consistently, see Fig. 4.7. At this point, we can focus our attention on another process where problems may occur. If there is no other process, we can continue to work on the first process to further reduce variability.

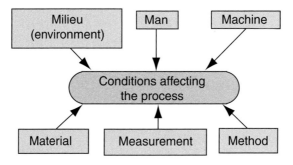

Figure 4.6 Main conditions affecting a process

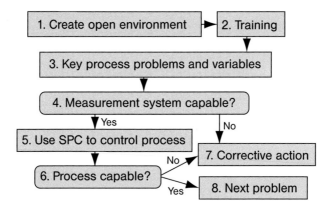

Figure 4.7 Flow diagram of implementation of SPC

4.3 Basic Statistical Concepts

4.3.1 Causes of Variability

All processes are subject to variability. Shewhart distinguished two basic causes of variability: *common causes* and *assignable causes*. A common cause is a source of variation that is inherent or natural to the process. They are also called *random causes, chance causes,* and *natural causes*. Examples of common causes in injection molding are ambient temperature, relative humidity, tooling adjustment, barrel temperature, among others. An assignable cause is a source of variation that is not inherent to the process but has an identifiable reason. They are also called *special causes, sporadic causes, chaotic causes,* and *unnatural causes*. Examples of assignable causes in injection molding are barrel heater burnout, unusual screw and barrel wear, sticking nozzle valve, contamination in raw material, and so on.

The effect of a common cause is typically slight. No major part of the total variation can be traced to a single common cause. Common causes influence all data in a similar manner. They are stable and the collective pattern of common causes is predictable.

The effect of an assignable cause can be strong. Often a major part of the total variation can be attributed to a single assignable cause. Assignable causes influence some or all data in a dissimilar manner. They are not stable and the effect of assignable causes is unpredictable. Table 4.2 summarizes the characteristics of common and assignable causes.

If only common causes of variation are present, the output of a process forms a distribution pattern that is stable over time and predictable, see Fig. 4.8.

If assignable causes of variation are present, the output of a process will not be stable over time and is not predictable, see Fig. 4.9.

Table 4.2 Characteristics of Common Causes and Assignable Causes

	Common cause	Assignable cause
Nature	Inherent or natural to the process Predictable Stable	Unnatural Unpredictable Unstable
Typical identity	Many small sources	One or just a few major sources
Effect	Slight, although in some cases can be strong	In most cases a strong effect
Improvement action	Reduce common cause variability	Eliminate
Improvement tools	Multi-vary charts, components search, paired comparison, design of experiments, etc.	Statistical process control
Responsibility	Management	Operator/supervisor

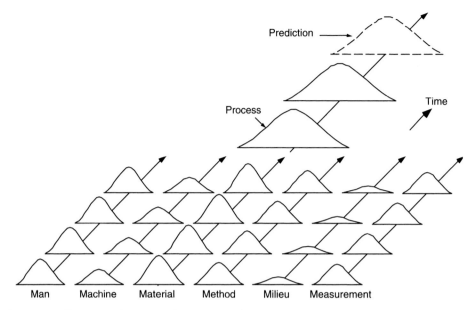

Figure 4.8 Illustration of a process that is in control

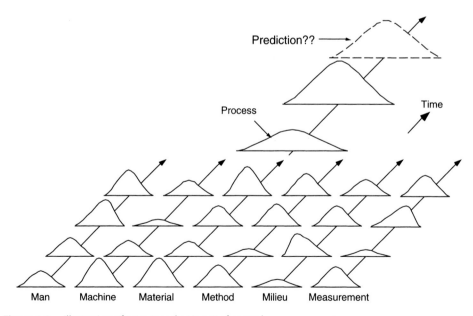

Figure 4.9 Illustration of a process that is out of control

When we want to reduce process variability we usually focus first on variability due to assignable causes. There are a number of good reasons for this. Assignable causes often have a strong effect on total variability and they are usually easier to eliminate.

4.3.2 Basic Statistical Terms

When we collect data on a process, we pull *samples* from the process. A sample is one or several individual pieces or measurements collected for analysis. Samples are usually pulled in small groups called *subgroups*. Subgroups that are collected in such a way that little variation can be expected between parts within the group, such as consecutive parts off an injection molding machine, are considered *rational subgroups*. They are called rational because the way the subgroup has been chosen, hopefully, makes a certain amount of sense. Subgroups typically contain five to ten individual pieces or measurements.

Subgroups, of course, are part of the total number of parts that are produced. All of the parts we manufacture make up a *population* or *universe*. By analyzing the variation of data in subgroups, we try to determine the variation in the population. A principal tool in the analysis is the control chart. Figure 4.10 shows the hierarchy of individual measurements, subgroups, and population; the subgroups in this example consist of five individual measurements each.

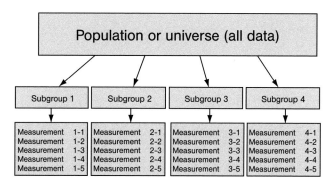

Figure 4.10 Hierarchy of individual measurements, subgroups, and population

4.3.3 Mean, Median, and Mode

In most processes we try to manufacture parts to a certain target dimension. This target dimension is called the *nominal*. Because of the natural variation in processes, the actual dimensions will not be exactly at nominal. Some will be higher and some will be lower. Hopefully, all dimensions will be close to the nominal. This tendency to group around a certain dimension is called *central tendency*. The most common measure of central tendency is the *mean* or *average*. It is the sum of the observations divided by the number of observations and is usually represented by the symbol $\bar{x}$. If we have a number of n observations with values $x_1, x_2, x_3, \ldots, x_n$, the mean is calculated as follows:

$$\bar{x} = \frac{x_1 + x_2 + x_3 + \ldots + x_n}{n} \tag{4.1}$$

For example, we have five measurements of the thickness of molded caps:

$x_1 = 0.0203$ in, $x_2 = 0.0208$ in, $x_3 = 0.0197$ in, $x_4 = 0.0198$ in, and $x_5 = 0.0199$ in

For this example the mean becomes:

$$\bar{x} = \frac{0.0203 + 0.0208 + 0.0197 + 0.0198 + 0.0199}{5} = 0.02010 \text{ in}$$

$\bar{x}$ is often termed the first moment of the distribution; it is analogous to the center of gravity. Another measure of central tendency is the *median*. It is the middle measurement with the numbers arranged in order of size. It is usually represented by $\tilde{x}$ (x-tilde), sometimes by the symbol Mi. For an even set of numbers, it is the average of the two middle numbers. If we arrange the numbers in the previous example in order of size we get:

0.0197 in

0.0198 in

0.0199 in ← median

0.0203 in

0.0208 in

In this example the median is 0.199 in.

A third measure of central tendency is the mode. It is the most frequently occurring value in a set of data. It can be represented by the symbol $\hat{x}$ (x-caret).

4.3.4 Range, Variante, and Standard Deviation

Mean, median, and mode are measures of central tendency. In addition to the central tendency, we also need to know about the spread of the data. Another term used for spread is dispersion. Commonly used measures for spread or dispersion are range, variance, and the standard deviation. The *range* is the difference between the largest and smallest measurement. The symbol for range is R; it is calculated from:

$$R = x_{max} - x_{min} \tag{4.2}$$

Variance is the mean of the squared deviations from the arithmetic mean. It is a kind of average that reflects the distance of individual measurements from the mean. The symbol for variance is v; it is calculated from:

$$v = \frac{(x_1 - \bar{x})^2 + (x_2 - \bar{x})^2 + (x_3 - \bar{x})^2 + \ldots + (x_n - \bar{x})^2}{n-1} = \frac{SS}{DF} \quad (4.3)$$

where n is the number of measurements, SS is the sum of squared deviations from the mean, and DF is degrees of freedom.

Variance is also called the second moment of the distribution. It is analogous to the moment of inertia. We will calculate the variance for the previous example with the measurements written as whole numbers:

$x_1 = 203$, $x_2 = 208$, $x_3 = 197$, $x_4 = 198$, and $x_5 = 199$

The mean was determined to be $\bar{x} = 201$. The variance now becomes:

$$v = \frac{(203-201)^2 + (208-201)^2 + (197-201)^2 + (198-201)^2 + (199-201)^2}{4} = 20.5$$

The *standard deviation* is simply the square root of the variance. The symbol for standard deviation is s; it is calculated as follows:

$$s = \sqrt{\frac{(x_1 - \bar{x})^2 + (x_2 - \bar{x})^2 + (x_3 - \bar{x})^2 + \ldots + (x_n - \bar{x})^2}{n-1}} \quad (4.4)$$

Of course, we can also write the much shorter expression:

$$s = \sqrt{v} \quad (4.5)$$

Thus the standard deviation for the example is:

$s = \sqrt{20.5} = 4.53$

Another measure of dispersion that is used in some cases is the coefficient of variation or *CV*. The *CV* is the standard deviation divided by the mean:

$CV = s/\bar{x}$

The coefficient of variation is frequently used in the analysis of mixing. The inverse of the coefficient is the signal to noise ratio, which is used in the Taguchi method; see Chapter 9.

The variance and standard deviation calculated so far are measures of the spread of the data in a sample. When we determine the variance and standard deviation of the entire population, we substitute the population mean for the sample mean $\bar{x}$. The population mean is usually represented by the Greek symbol μ (mu). The population standard deviation is typically represented by Greek symbol σ (sigma), while the sample standard deviation by symbol s.

If the population has a manageable size, such as 80 or 125, the population mean and standard deviation can be determined. However, when the population is very large, such as 450,000, it is often not practical to determine the mean and standard deviation of the population. In this case, the population mean is often approximated by the mean of the averages of several subgroups.

Variance has the property of additivity. This means that the total variance is equal to the sum of the variances of its parts. Standard deviation does not have the property of additivity. By analyzing the variances of different factors, it is possible to identify how these factors contribute to the total variance. This is known as the analysis of variances or ANOVA.

The Greek characters are used to represent characteristics of the population, while Roman characters (regular alphabet) are used for characteristics of the samples. The standard deviation is a useful measure of the spread of data when the data follow a normal distribution. The normal distribution is very important in SPC; distribution patterns will be covered next.

4.3.5 Plotting Distribution Patterns

All measurements are subject to variation. Their pattern can be graphically represented by distribution curves; these can be:

1. Histograms
2. Frequency polygons
3. Frequency curves

A histogram is basically a bar chart with the height of each bar indicating how many data occur within a certain interval; the width of the bar equals the width of the interval. To plot the distribution pattern, the following guidelines can be used:

1. Count the number of measurements (N).
2. Determine the number of intervals (K) from Table 4.3.
3. Determine the data range (R), that is, highest minus lowest value.
4. Estimate interval size by R/K.
5. Round estimate interval to a convenient number.
6. Determine interval boundaries.
7. Determine one-half number accuracy.
8. Adjust interval boundaries.
9. Tabulate data.

As an example we will look at measurements of wall thickness in injection molded caps expressed in inches; see Table 4.4.

4.3 Basic Statistical Concepts

Table 4.3 Guide for Number of Intervals in Histograms

Number of measurements (N)	Number of intervals (K)
30 to 50	5 to 7
51 to 100	6 to 10
101 to 250	7 to 12
Over 250	10 to 20

Table 4.4 Data of Wall Thickness of Caps

0.0207	0.0198	0.0208	0.0199	0.0210	0.0197
0.0189	0.0192	0.0198	0.0197	0.0192	0.0191
0.0204	0.0201	0.0190	0.0203	0.0196	0.0194
0.0213	0.0208	0.0201	0.0207	0.0204	0.0206
0.0201	0.0197	0.0217	0.0201	0.0211	0.0204

To make the data easier to handle, we will write the data in whole numbers; see Table 4.5.

In this example the number of measurements $N = 30$. From Table 4.3 we select $K = 6$. The highest value $x_{max} = 217$ and the lowest $x_{min} = 189$. Thus the range $R = 217 - 189 = 28$. The estimated interval size becomes $28/6 = 4.67$. We will round the interval size to 5. We can now set the interval boundaries to 185, 190, 195, 200, 205, 210, 215, and 220. The one-half number accuracy is 0.5 and the adjusted interval boundaries become: 185.5, 190.5, 195.5, 200.5, 205.5, 210.5, 215.5, and 220.5. When the data are whole numbers, the one-half number accuracy is always 0.5. The data can now be tabulated as shown in Table 4.6.

Table 4.5 Data of Wall Thickness of Caps

207	198	208	199	210	197
189	192	198	197	192	191
204	201	190	203	196	194
213	208	201	207	204	206
201	197	217	201	211	204

Table 4.6 Thickness Data Tabulated by Tallies

| 185–190** |
| 191–195**** |
| 196–200****** |
| 201–205******** |
| 206–210****** |
| 211–215** |
| 216–220* |

Another way of displaying the distribution of the data is by using a stem-and-leaf plot. In this plot the first two digits form the stem and the third digit forms the leaf.

In this stem-and-leaf plot, the intervals run from 185 to 189, 190 to 194, 195 to 199, and so on. In the second interval (190–194) there are five data: 192, 192, 191, 190, and 194. The five leaves representing the last digit, that is, 2, 2, 1, 0, and 4, indicate the five data in the second interval, with 19 being the stem.

The advantage of a stem-and-leaf plot is that it not only shows the distribution of the data, it also shows the actual values of the data (Fig. 4.11).

18 9
19 22104
19 8978767
20 41314114
20 78876
21 031
21 7

Figure 4.11 Data from Table 4.5 in stem-and-leaf plot

The histogram as a proper bar chart is shown in Fig. 4.12.

The frequency polygon plots the frequency in the center of the interval and connects the points by straight lines. This is shown in Fig. 4.13.

By drawing a smooth line through the points, we obtain the frequency distribution curve; see Fig. 4.14.

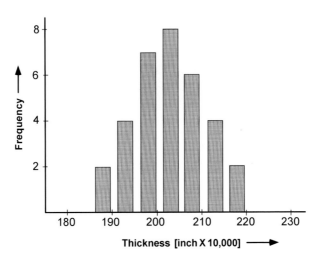

Figure 4.12 Bar chart of the wall thickness data

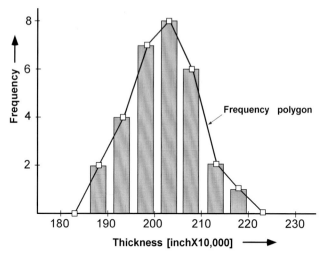

Figure 4.13 Wall thickness data shown as frequency polygon

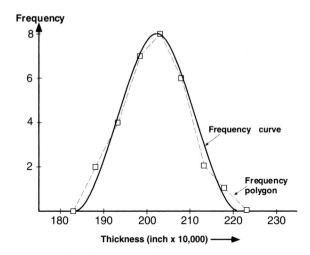

Figure 4.14 Smooth frequency distribution curve of wall thickness data

It should be noted that these guidelines for plotting histograms are not unique; many other guidelines are employed. The easiest way to plot a histogram is to use a software package that has the ability to draw histograms. Here, one only has to enter the data in a spread sheet format and click on the right pull-down menu and the histogram will appear on the screen within a few seconds. This is not only much easier and faster, it also produces more professional looking results and minimizes the chance of errors. The histogram in Fig. 4.15 was created using Minitab.

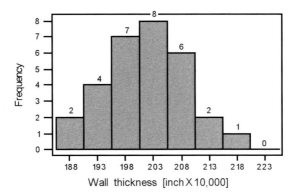

Figure 4.15 Histogram created using SPC software (Minitab)

4.3.6 Characteristics of a Frequency Distribution

Distributions can be described by their modality, variability, degree of skew, and degree of kurtosis. Some distributions have more than one point of concentration. These are said to be multimodal distributions. A distribution with a single point of concentration is called unimodal. Figure 4.16 shows a unimodal distribution on the left-hand side and a bimodal distribution on the right-hand side.

If one distribution is narrow and another broad, both using the same horizontal scale, it means that the first has less variability than the second. This is illustrated in Fig. 4.17.

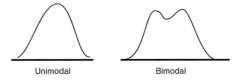

Figure 4.16 Unimodal and bimodal distributions

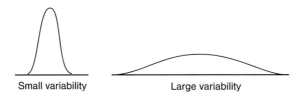

Figure 4.17 Narrow and wide distributions

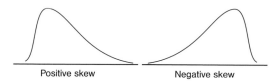

Figure 4.18 Illustrations of positive and negative skews

A third characteristic is the symmetry of the variation; is it symmetrical or lopsided? If it is lopsided, the distribution is said to be skewed. Figure 4.18 shows a distribution with positive skew on the left and a negatively skewed distribution on the right. A distribution that is symmetrical has no skewness.

The fourth characteristic deals with the relative concentration of data at the center and along the tails of the distribution. This characteristic is called kurtosis. If a distribution has a relatively high concentration in the middle and out on the tails, but relatively little in between, it has large kurtosis. If it is relatively flat in the middle and has thin tails, it has little kurtosis. This is illustrated in Fig. 4.19.

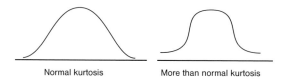

Figure 4.19 Illustrations of large and normal kurtosis

4.3.7 Different Distribution Patterns

Data can be distributed in a variety of different patterns. Some of the more important ones are:

1. The binomial distribution
2. The log normal distribution
3. The normal distribution

Other distributions are the Poisson and exponential distributions; these, however, are used less often in SPC.

The binomial distribution is used when the result of an inspection can only have one of two possible outcomes, such as good or bad. A product characteristic that can only be described by one of two possible conditions is called an *attribute*. This is as opposed to a *variable*, which is a product characteristic that can be described by one of many possible values. Attributes and variables will be covered in more detail later. The binomial distribution is useful in SPC when we deal with product attributes. When we are dealing with large samples, typically fifty or more, then the binomial distribution is very closely approximated by the normal distribution. This is true as long as the less probable of the two possible outcomes occurs at least four or five times in each sample.

The log normal distribution is a special type of normal distribution. Log is short for logarithm. We talk about a log normal distribution when a frequency curve of the logarithm of each of the individual points forms normal distribution curve. The log normal distribution is often used in the analysis of particle size distribution of solid materials, because many of them follow it closely.

The normal distribution is probably the most important one in SPC. It is also called Gaussian distribution. The distribution curve is bell-shaped; see Fig. 4.20.

The normal distribution has the following characteristics:

- It is symmetrical about the mean, therefore the mean, median, and mode are all equal.
- It slopes downward on both sides to infinity.
- 68.25% of all measurements lie between $\mu + 1\sigma$ and $\mu - 1\sigma$; see Fig. 4.20.
- 95.46% of all measurements lie between $\mu + 2\sigma$ and $\mu - 2\sigma$.
- 99.73% of all measurements lie between $\mu + 3\sigma$ and $\mu - 3\sigma$.

The equation for the bell-shaped normal distribution curve is presented in Appendix IV. The area under a particular portion of the curve can be found in charts, usually called z charts; see Appendix IV.

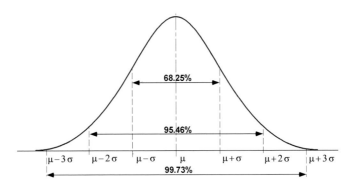

Figure 4.20 The normal or Gaussian distribution curve

4.3.8 The Central Limit Theorem

In using the normal distribution to determine how many measurements can be expected to be within certain values, we have to be sure that the data follow a normal distribution. The central limit theorem developed by Shewhart is very important in this respect. It states that sample averages (or $\bar{x}$) will follow a normal distribution as long as only common cause variations are present. This is true even if the individual measurements do not follow a normal distribution.

The standard deviation of the averages of subgroups, $\sigma_{\bar{x}}$, will be smaller than the standard deviation of the individual measurements, σ_x; see Fig 4.21.

The two standard deviations are related by the formula:

$$\sigma_{\bar{x}} = \frac{\sigma_x}{\sqrt{n}} \qquad (4.7)$$

where n is the number of data in the subgroup.

The charting of averages is an important element of control charts. The central limit theorem is an important principle upon which many control charts are based.

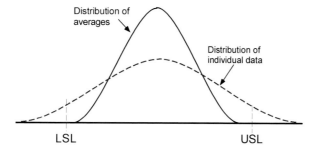

Figure 4.21 Comparison of the distribution of averages and the distribution of individual data

5 Data Collection, Data Analysis, and Problem Solving

5.1 Attributes Data Versus Variables Data

As discussed in the previous chapter, data can be divided into attributes data and variables data. Variables data are data that can be measured on a continuous scale. Attributes data are data that can have only one of two possible outcomes. A comparison between the two types of data is shown in Table 5.1.

Table 5.1 Comparison of Attributes to Variables Data

Variables data	Attributes data
Results from a measurement Example: 97.4 °F, 16.24 oz.	Observations counted Example: yes/no, present/absent
Employs a measurement tool • Scale • Thermometer • Micrometer	Employs visual inspection • Go/no go gauging • Pass/fail inspection
Continuous scale	Discrete scale

When collecting data on the product, it is important to measure meaningful data. These are primarily determined by the functional requirements of the product and the customer requirements. In a continuous process, such as extrusion, or in a cyclic process, such as injection molding, it is most efficient to measure product characteristics *on-line* (the buzzword that is used for this is real-time monitoring). This allows a rapid detection of problems and quick response before a large amount of out-of-spec product is produced. Measurements in injection molding and extrusion will be discussed in Chapter 6.

5.2 Important Aspects of Data Collection

There are a number of important considerations related to data collection. They are measurement system evaluation, sampling methods, data handling, and process control systems. The analysis of a process is not meaningful unless the measuring instruments used to collect the data are both accurate and repeatable. Because of the importance of measurements, this subject will be covered in more detail in Chapter 6.

The accuracy of the analysis of a process is also dependent on the appropriateness of the sampling methods used. Data handling is an issue that can be important in data acquisition systems. These systems often sample at high rates and use some kind of averaging in the reporting of the data. This averaging, however, can mask certain process variations depending on the frequency of the variation and the averaging procedure. When using an electronic data acquisition system (EDAS), it is important to know how the data are sampled and analyzed internally.

5.3 Diagrams for Problem Solving

Various diagrams can be used to help in problem solving. *Dr. Kaoru Ishikawa* developed several such diagrams called cause and effect diagrams. As a result, they are often called Ishikawa diagrams. These diagrams are generally shaped as a fish skeleton; therefore, they are sometimes referred to as fishbone diagrams. Three important cause and effect diagrams are:

- Cause Listing Diagram: A listing of the possible causes of a problem
- Variation Analysis Diagram: A diagram used to analyze causes of variability in a process
- Process Analysis Diagram: A flow diagram to study the various steps in a process

These three diagrams look quite similar. The last two were derived from the first, the basic cause listing diagram.

5.3.1 Cause Listing Diagram

In a cause listing diagram, the problem is put at the far right side of the page, as shown in Fig. 5.1.

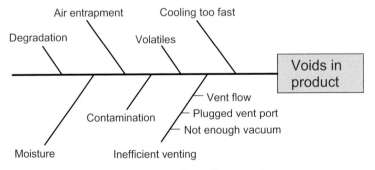

Figure 5.1 Example of a cause listing diagram for voids in a product

The diagram shown lists a number of possible causes for voids in a product. It is important to make this list as complete as possible. Therefore, it is important to solicit input from all the people associated with any aspect of the problem. A brainstorming session is a good way to come up with a large number of possible causes.

5.3.2 Variation Analysis Diagram

A variation analysis diagram is constructed the same way as a cause listing diagram. The problem is listed on the far right. The main conditions affecting the process are listed: man, method, machine, material, measurement, and milieu. Next, all the details that can contribute to the variability are listed under the main categories. An example is shown in Fig. 5.2, which lists possible causes for dimensional variation in an injection molded product.

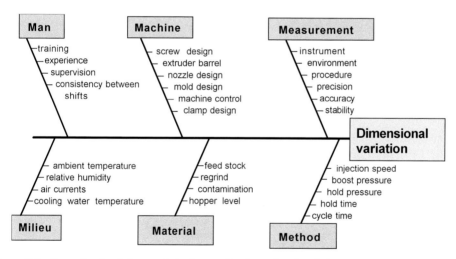

Figure 5.2 Example of variation analysis diagram for dimensional variation

The next step is to list details that can contribute to variability by adding arrows to the branches and labeling them. An example is shown in Fig. 5.3 for the main branch with material related sources of dimensional variation.

It is clear that several diagrams may be necessary to list all possible sources of variation.

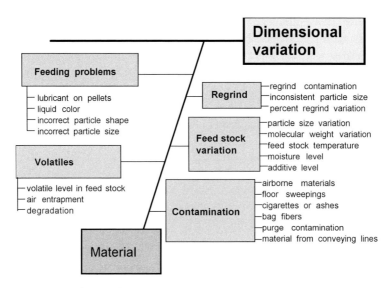

Figure 5.3 Example of detail of variation analysis diagram

5.3.3 Process Analysis Diagram

In a process analysis diagram each process step is labeled and connected by a line going from left to right; see Fig. 5.4.

At each process step everything that can influence the product quality is added by branches. Figure 5.4 shows a process analysis diagram for a typical injection molding operation. All the process steps are listed with factors that can potentially cause the problem.

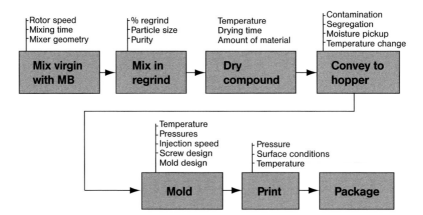

Figure 5.4 Example of process analysis diagram for injection molding process

The benefit of all these diagrams is that all critical factors have to be determined and put on paper. This allows a more systematic approach to problem solving. The drawback is, of course, that the diagrams are only as good as the people putting them together. It is important, therefore, to solicit input from as many sources as is reasonably possible. To summarize the benefits:

- All factors can be determined.
- A systematic approach to problem solving can be taken.
- The diagram provides a good record for future reference and may make problem solving easier and quicker in the future.
- When the analysis has been completed, we can be reasonably sure that all factors have been accounted for.

5.4 Pareto Diagrams

Pareto diagrams are a useful tool in separating the "critical few" from the "trivial many." In most processes, the majority of the quality problems result from relatively few causes. Pareto diagrams show how many problems are caused by the various causes. The most frequently occurring problem is listed first, the second most frequent problem next, and so on. The Pareto diagram is basically a bar chart with each problem represented by a different bar; the height of the bar represents the frequency of the problem.

Pareto diagrams can be used with both variables and attributes data; however, they are most often used with attributes data. The data are usually expressed in percentages or number of occurrences. We can take an example of defects in injection molded parts over a 24 hour time period as determined by visual inspection; the data in Table 5.2 are obtained.

To construct the Pareto diagram, it is convenient to rearrange the data as shown in Table 5.3.

Table 5.2 Table Showing Defects in Injection Molded Parts

Defect	Number	Percentage	Ranking
Splay	4	2.67%	5
Black specks	7	4.67%	4
Voids	78	52.00%	1
Sink marks	23	15.33%	2
Flash	12	8.00%	3
Other	26	17.33%	
Total	150	100.00%	

Table 5.3 Table Showing Defects in Injection Molded Parts

Defect	Number	Percentage	Ranking
1. Voids	78	52.00%	1
2. Other	26	17.33%	2
3. Sink marks	23	15.33%	3
4. Flash	12	8.00%	4
5. Black specks	7	4.67%	5
6. Splay	4	2.67%	
Total	150	100.00%	

Figure 5.5 is a Pareto chart showing the number of defects in each category on the right vertical axis and the percentage defects on the left vertical axis. Also shown in Fig. 5.5 is a line showing the cumulative percentage.

The main benefits of Pareto diagrams are:

- They uncover the few most important causes of problems.
- They can clearly demonstrate the results of improvements.

When using Pareto diagrams, it is important to use numbers of defects on one of the vertical axes; Fig. 5.5 shows the number of defects on the right-hand side vertical axis. This will clearly show the effect of improvement efforts. The percentages will not clearly show the effect of improvements; percentages are shown on the left-hand vertical axis. Pareto diagrams, together with cause and effect diagrams, are useful in determining and directing the most efficient improvement efforts and in following the results of these efforts.

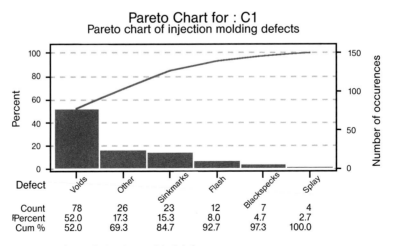

Figure 5.5 Pareto chart of injection molded defects

5.5 Histograms

When data are collected for a process, we want to know how much variation occurs in the data. More specifically, we want to know whether the data are centered around a single value, how wide the distribution is, whether the distribution resembles a normal distribution, and so on. A histogram can answer these questions. The construction and interpretation of histograms was discussed earlier in Section 4.3.5.

5.6 Scatter Diagrams and Correlation Tables

When we have determined the causes contributing to a problem, we need to find out how strong an effect each cause has. One method of determining the cause and effect relationship is the scatter diagram; another is the correlation table. A scatter diagram is simply a graph showing the cause on the horizontal scale and the effect on the vertical scale. Figure 5.6 shows the effect of extruder barrel temperatures on product brittleness.

From Fig. 5.6 it is clear that increasing the extruder barrel temperatures increases product brittleness. Another way of uncovering the cause and effect relationship is by using the correlation table. This table is useful with a large number of data and when many data have the same value. An example is shown in Table 5.4. In this correlation table, the effect of mold temperature on the warpage of the molded product is shown. The table shows that in this case the warpage reduces with increasing mold temperature. Also listed in the table along the top and the right-hand side is the number of data in each column and row.

When we plot a scatter diagram, a number of situations can occur. When the effect increases with the cause, we describe this as a positive correlation; see Fig. 5.7.

When there is little scatter in the data, we will call this a strong positive correlation; see Fig. 5.7, left. When there is substantial scatter in the data, this is called a weak positive

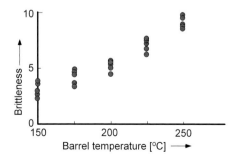

Figure 5.6 Scatter diagram of product brittleness related to barrel temperature

5 Data Collection, Data Analysis, and Problem Solving

Table 5.4 Correlation of Mold Temperature Versus Warpage

Number of data →	10	10	10	10	10	50
2.0						0
1.8	*					1
1.6	**					2
1.4	*****	**				7
1.2	**	****	**			8
1.0		***	****	**	*	10
0.8		*	***	***	****	11
0.6			*	****	***	8
0.4				*	**	3
0.2						0
$\Delta H \uparrow$	100	110	120	130	140	150
(mm)	Mold Temperature (°F) →					

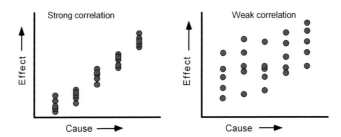

Figure 5.7 Example of a positive correlation, strong (left) and weak (right)

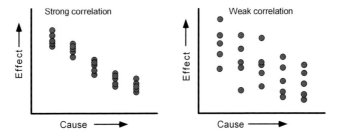

Figure 5.8 Examples of a negative correlation, strong (left) and weak (right)

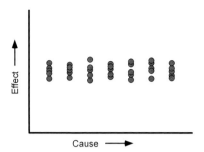

Figure 5.9 Example of a case with no correlation

correlation; see Fig. 5.7, right. When the effect decreases with the cause, we have a negative correlation, see Fig. 5.8.

Again, we can have a strong negative correlation; (see Fig. 5.8, left) or a weak negative correlation (see Fig. 5.8, right). Finally, we can have a situation where there is no correlation at all; see Fig. 5.9.

It is possible to determine a value to describe the correlation between two factors. This is called a correlation coefficient (r). There are several methods of calculating correlation coefficients. In all methods the value of r ranges from -1 to $+1$. A value of -1 indicates a perfect negative correlation, a value of $+1$ shows a perfect positive correlation, and a zero indicates no correlation.

5.7 Recording Data

There are three common ways to record data for process control. They are check sheets, portable data collectors, and fixed station data acquisition systems. Each method has advantages and disadvantages.

5.7.1 Check Sheets

Check sheets are easy to make and the most flexible way to record data. They should be designed to allow efficient data collection with minimum chance of error. Errors occur more often when copying or reading data along a horizontal line than down a vertical column. Therefore, columns should be used for recording groups of data rather than rows. Pertinent information should be included in the check sheet, such as the name of the data collector, the time and date, shift, machine, purpose of the study, and so on. There should also be room for notes on the check sheet.

Check sheets can be used both for variables data and for attributes data. Variables data can be recorded in two ways, in columns with one column for each type of measurement or in a histogram when only one type of measurement is recorded on a check sheet. A check sheet for variables with columns is shown in Fig. 5.10.

Product: Connector tube
Injection machine: number 4 Material: HDPE, Lot number 135
Molded: 7-14-95, work order 17451
Specifications: L = 12.000" ± 0.020"
ID = 0.200" + 0.010"
OD = 0.300"+ 0.010"
Date/time: 7-15-95, 7:15 a.m.
Auditor: Jeff Anderson
Remarks:

	L	OD	ID
	12.002	0.298	0.199
	12.007	0.299	0.203
	11.991	0.306	0.192
	11.998	0.291	0.205
	11.987	0.301	0.198
	12.012	0.303	0.204
Average	11.9995	0.2997	0.2002
Range	0.020	0.008	0.013

Figure 5.10 Check sheet for variables data with columns

This figure shows a check sheet to monitor molded connector tubes with three dimensions being measured; length, inside diameter, and outside diameter. A check sheet for variables in histogram form is shown in Fig. 5.11. The item being measured is the shrinkage of a molded panel.

Check sheets for attributes data can be one of three types (see [2]): defect-by-item, defect-by-location, and defect-by-cause. An example of a defect-by-item check sheet is shown in Fig. 5.12.

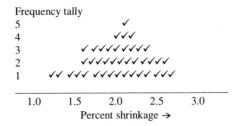

Figure 5.11 Check sheet for shrinkage data of molded panel

Defect tally sheet

Date/time: 3-26-91/2:40 p.m.
Material: 160 housing for customer 2207, order XB701
Inspection: F. Fernandez
Produced on: Line 5, 4-17-96
Remarks: Standard operating conditions

Defect	Number	Total
Voids	✓✓	2
Sink Marks	✓✓✓✓✓	5
Brown Spots	✓✓✓✓✓ ✓✓✓✓✓ ✓✓✓✓✓ ✓✓✓✓✓ ✓✓✓✓✓ ✓✓✓✓✓ ✓✓✓✓✓ ✓✓✓✓✓	40
Splay	✓✓✓✓✓ ✓✓✓✓✓ ✓	11
Flash	✓✓✓✓✓ ✓	6
Other	✓✓✓✓✓ ✓✓✓✓✓ ✓✓✓✓	14
	Grand Total	78

Figure 5.12 Defects-by-item check sheet for defects in molded parts

5.7.2 Portable Data Collectors/Machine Analyzers

Portable data collectors (PDCs) are similar to check sheets in that they can be easily moved around. In injection molding, these devices are often referred to as portable machine analyzers (PMAs). They have some important advantages:

- They can record data in computer-readable form.
- They can take data directly from electronic sensors and gages. This makes PMAs fast and minimizes errors.
- Data can be analyzed internally to yield information on mean, range, maximum value, minimum value, standard deviation, and so on.
- They are available with limit checking; the PMA can give an alarm when data just taken are out of specification.

PMAs can collect variables as well as attributes data. The use of PMAs in injection molding has increased considerably as the prices have come down to levels that are affordable even for small molding operations [31]. Several PMAs are presently available for less than $10,000. Typical analog inputs from transducers include clamp and injection pressure, screw position and speed, and several process temperatures. Digital inputs from controllers can show events such as injection forward and mold open.

Most PMAs offer the user some flexibility in assigning inputs to the data acquisition channels; this even extends to auxiliary equipment, such as dryers, and external signals, such as plant ambient temperature and relative humidity. The most important analog inputs are cavity pressure, fill time, screw velocity, and screw position. Important digital inputs are injection forward, holding, recovery, and mold open. Other parameters to be monitored may depend on the specifics of the operation. For instance, in injection molding of PVC, useful parameters tend to be cooling water temperature and flow rate. A typical system can monitor four to six channels. Many users have found that systems using many more channels tend to become overwhelming and confusing. However, for advanced process research, systems capable of handling 32 channels or more may be required.

PMAs are typically used to analyze the health of an injection molding process. Being able to display the cavity pressure-time curve is like having a fingerprint of the part. PMAs are like having an electrocardiogram of an injection molding process. Some companies use PMAs during process development and start-up to quickly find the optimum process conditions for new molds. Other companies use PMA tests to validate the repeatability of new machine performance before final payment.

Machine analysis using PMAs can lead to surprising findings. One company found that cushion control on an injection molding machine made no difference in machine performance. As a result, this company will order new machines without this feature. Another company found that the supervisory control feature of a machine vendor actually deteriorated repeatability. As a result, they turned off the supervisory control.

One PMA from Nicollet Process Engineering, called the Personal Computer Machine Analyzer (PCA), offers a logic analyzer, a profile analyzer, a chart recorder, and a volt meter in a single system. The PCA can pinpoint problems quickly with four diagnostic functions [34]. Another application is the development of an operational thumbprint for a new injection molding machine by original equipment machinery suppliers (OEMs). Signatures are stored electronically, allowing operators to compare current machine readings to the baseline thumbprint from the factory. The PCA can evaluate more than sixty parameters, including cycle time, clamp close time, ram velocity, cavity pressure, and mold temperature. It has up to sixteen analog and sixteen digital channels that can be used to analyze any type of machine with digital outputs. The PCA can take about two hundred samples per second. Appendix VI shows a listing of some monitoring systems for injection molding.

5.7.3 Fixed Station Data Acquisition Systems

As the name implies, fixed station data acquisition systems (DAS) are fixed to one location, either because of size or because the wiring makes it very difficult to move the unit. A fixed station DAS can have a wide range of capabilities. A simple DAS may record data from only one injection molding machine, that is, it may be a dedicated DAS. A number of machine suppliers now offer injection molding machines with integrated data acquisition and SPC capability; some examples are listed in Appendix III.2.

A more sophisticated DAS may be capable of recording data from various sources, analyzing the data, presenting control charts, showing trend plots for different variables, and so on. These systems are often referred to as plant-wide monitoring systems. At least one manufacturer of injection molding machines supplies a central computer system that allows the user to view the operation of plant equipment and change the operating parameters on any selected equipment. A fixed station DAS can take many different forms depending on the application. A good DAS can be a very valuable tool in improving process control and also in problem solving. For application to injection molding the following capabilities are useful:

- *Monitoring of Many Variables:* A typical molding process requires about twenty process parameters to be monitored.
- *Data on Slowly Changing Variables:* These should be taken at least once a second. For rapidly changing variables, such as melt pressure and hydraulic pressure, data should be collected at higher frequency, typically 100 points/second. Some high-end systems sample up to 100,000 points/second.
- *Trending:* The ability to display the variation of one or more process parameters over a particular time period. It is useful if the scales in these displays are adjustable. This capability is extremely helpful in troubleshooting and problem solving.
- *Determination of Statistical Parameters:* These might be parameters on process and product, such as mean, standard deviation, control limits, and so on.
- *Alarms for Out-of-Spec Data:* indications of assignable causes of variation.
- *Recipes:* The system might have the capability to store important process parameters for different products. This allows previous process conditions to be reproduced quickly and reliably.
- *Production Summaries:* The system might be able to follow the amount of material being produced and present summaries per shift, per day, per week, and so on. This is a useful management tool to analyze productivity on different production lines.

The following features have to be considered:

- *User-Friendliness:* The system should be easy to use, intuitive, and should not require very long training for operators to become accustomed to the system.
- *Accessibility:* Is access from a remote computer possible? With remote access, integration into a plant-wide information control system is possible.
- *Connectivity:* Can the DAS software work together with other software packages?
- *Upgradability:* Can new upgraded versions of software and/or hardware be readily implemented?

- *Cost:* An analysis should be made to determine whether the cost savings in terms of improved production efficiency and quality outweigh the cost of a data acquisition system.

The different methods of data collection each have their advantages and disadvantages. Check sheets are slow, prone to human errors and transcription errors; however, they are also inexpensive, flexible, efficient for small amounts of data, and easy to use. Fixed station data acquisition systems are fast, not prone to human errors and transcription errors, can handle large amounts of data; however, they tend to be expensive and more difficult to use. Thus check sheets and fixed station data acquisition systems have complementary areas of applications. Portable data collectors fall in between check sheets and fixed station data acquisition systems. Table 5.5 shows a comparison of the different data recording methods.

For plant-wide monitoring systems, the injection molding process can be integrated with upstream and downstream operations. Bar coding can be used to achieve parts traceability and can be integrated with automatic product tracking and warehousing. With these systems, accurate data can be collected for job costing, such as cycle times, yields, efficiencies, scrap, labor allocation, and downtime — essential information for efficient plant management. Functions that can be included in plant-wide monitoring systems are preventive maintenance, production scheduling, inventory, production reporting, order entry, job histories, real-time alarms, quality control, SPC, and others.

Some plant-wide monitoring systems allow changes to be made by remote control from a central terminal, in such parameters as screw speed, transfer point, backpressure, and so on. Obviously, the results from these changes have to be monitored carefully, because they can make the process not only better, but also worse. Some suppliers of plant-wide monitoring systems are listed in Appendix IIIC.

Table 5.5 Comparison of Data Recording Methods

	Check sheets	Portable data collectors	Fixed station DAS
Speed	Slow	Fair	Fast
Human errors	Substantial	Small	Very small
Transcription errors	Substantial	No	No
Efficient for:	Small amount of data	Large amount of data	Very large amount of data
Cost	Minor	Fair	Substantial
Flexibility	Very good	Fair	Good
User skill required	Low	Medium	High

5.8 Sampling

A population is the collection of all individual items in a designated group. Generally, populations are very large, making it impractical to collect data on each individual item. This is why we use samples. Sampling involves taking a small representative group from the population. For a sample to be representative, every item must have an equal chance of being selected for the sample. Sometimes random number tables are used so that the samples chosen are truly random. These samples are then evaluated for the characteristics that we are interested in, such as surface finish, length, diameter, and so on. The results from the samples are then translated back to estimate the characteristics of the population.

For most process control applications it is recommended that the sample size be kept small, often as few as three or four items per sample and certainly not more than twenty. Rational subgroups or samples are collections of individual measurements or observations subject only to common cause variations. A system that is influenced only by common causes of variation is called a *common cause system*. Samples should be chosen to maximize the chance that the individual items within the sample are subject to a common cause system. Based on the following considerations, the sample size should be:

1. Small to minimize chances of special causes within the sample
2. Large to ensure good characterization of the actual process
3. Large to have the sample averages more closely follow a normal distribution
4. Large to provide sensitivity for the detection of process shifts
5. Small to keep the data acquisition cost reasonable

A sample size of four or five often provides a good balance of the various considerations. It is often tempting to combine the output of several processes that are assumed to be identical, for example, two equal size injection molding machines side by side running the same product. However, such nonrational sampling can cause a number of problems, such as stratification and mixing. With stratification, as shown in the next chapter, the $\bar{x}$ control chart shows wide limits relative to the plotted points. This is due to the fact that the variability within the subgroup is subject to more than just common cause variation. It is important to monitor a *single* process only.

6 Measurement

6.1 Introduction

When a process is analyzed, the observed variation is often assumed to be the same as the actual variation. There is, however, an important component in the observed variation, which is the variation due to the measurement. When the measurement variation is large, the observed variation can be substantially different from the true variation. In the extreme situation, the measurement variation can be so large that it overshadows the actual variation. Obviously, in this case accurate process analysis is not possible. Therefore, accurate process analysis requires a determination of the variation due to measurement to make sure that this is only a small part of the observed variation.

Measuring is the process of quantification, comparing an unknown magnitude to a known magnitude. The science of measurement is called *metrology*. To quantify data on parts or processes, defined standard units must be used. These are called *units of measure*. Several systems of international units have been developed. The most important one is the SI system, which is an acronym for the French words: Système International d'Unités (International System of Units). The SI system uses meter as the basic measurement of length, newton for force, second for time, and kilogram for mass. Most industrialized countries have adopted the SI system with the exception of the United States. In the US, the English system is still widely used; since England has converted to the SI system, it might be more appropriate to talk about the American system. This uses the basic measurements of foot for length, pound for force, and second for time. However, increasing international contacts are bringing about an increased use of the SI system in the US.

Any practical system of measurement standards should be based on units that are unchangeable. Primary reference standards are maintained at the national bureau of standards. In the US this bureau is called the National Institute of Standards and Technology (NIST). The standards consist of copies of the international kilogram and measuring systems that can verify the units and subunits of the defined standards. The primary reference standards are used for calibration of measuring instruments. Since the primary reference standards cannot be used for calibration of all instruments, secondary and tertiary standards were developed. These allow transfer of the primary standards for calibration of instruments in laboratories and manufacturing areas. Measuring instruments used by operators, technicians, process engineers, and so on, are calibrated against a set of working standards. The working standards are referred back to the primary standards through transfer standards. A precision of about five to one is required to transfer from one standard to the next. When an instrument is calibrated such that it can be related to a primary standard, the calibration is said to have traceability. The hierarchy of standards is shown in Fig. 6.1.

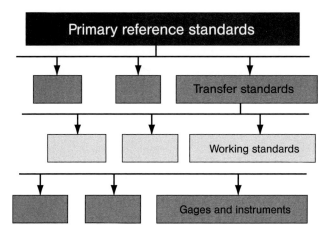

Figure 6.1 The hierarchy of measurement standards

6.2 Basic Concepts

Various terms are used to describe the characteristics of measuring instruments. Two important terms are accuracy and precision. These terms are often used interchangeably. However, they have distinctly different meanings in describing measuring instruments. *Accuracy* is the extent to which the average of many repetitive measurements on a single item agree with the true value. The difference between good accuracy and poor accuracy is illustrated in Fig 6.2.

An accurate instrument may or may not be precise. *Precision* is the extent to which repetitive measurements on a single unit agree. The difference between good precision and poor precision is illustrated in Fig. 6.3. It should be noted that the true value is not shown in Fig. 6.3, because the true value is not needed to evaluate precision.

The term bias is related to accuracy. *Bias* is the difference between the true value and the average of repetitive measurements made by a measurement process. When the bias is large, the measuring instrument is said to be out of calibration. The *sensitivity* of a measuring

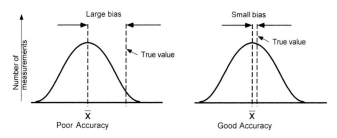

Figure 6.2 Illustrations of poor and good accuracy

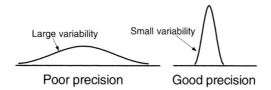

Figure 6.3 Illustrations of poor and good precision

instrument is the smallest change in the measured quantity that the instrument is capable of detecting. *Resolution* of a measuring instrument is the smallest unit of measure that an instrument is capable of indicating. Sensitivity and resolution are related terms; however, they have different meanings. *Sensitivity* describes the ability of a measuring instrument to *detect or sense* small changes, while *resolution* describes the ability to *indicate or display* small changes.

Stability refers to the difference in the average of at least two sets of measurements obtained with the same measuring device on a single unit taken at different times. This is illustrated in Fig. 6.4.

Stability is often expressed as a percentage of full scale. An ideal instrument has an absolutely linear relationship between the measured quantity and the measured output. The deviation from linear or straight line behavior is described by the term nonlinearity. *Nonlinearity* is the largest deviation of the actual measurements from the straight line characteristic divided by the full scale; see Fig. 6.5.

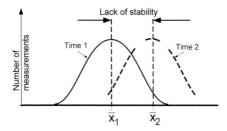

Figure 6.4 Illustration of lack of stability

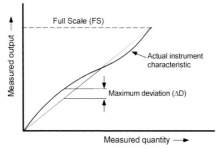

Figure 6.5 Illustration of non-linearity

6.3 Measurement Error

Measurement error is the difference between the measured value and the true value. It can be due to a problem of accuracy and/or precision. Various sources of error play a role in most instruments: nonlinearity, deformation, hysteresis, poor electrical connections, and sensitivity to environmental factors such as temperature, humidity, magnetic fields, electrical fields, and so on. Another important factor can be operator error. When two operators measure the same product with the same measuring instrument, there will be differences in the recorded values. Differences can also occur when different test procedures are used to measure the same product. Transposition errors can occur when data have to be read and recorded by the operator. If the data are not recorded legibly, transcription errors can occur later when the data are analyzed.

To keep measurement error very small it is recommended that the ratio of product tolerance (upper spec limit minus the lower spec limit) to the precision of the measurement instrument be at least 5:1 and preferably 10:1. This is illustrated in Fig. 6.6.

The observed variation is a combination of true product or process variation and measurement variation. This relationship can be expressed as:

$$\sigma_{obs} = \sqrt{\sigma_{meas}^2 + \sigma_{true}^2} \qquad (6.1)$$

In this expression σ_{obs} is the standard deviation of the observed data, σ_{meas} the standard deviation due to measurement, and σ_{true} the actual standard deviation of the product.

To minimize measuring error, instruments should be calibrated regularly. Calibration control should provide for periodic audits on instruments to check their accuracy, precision, and general condition. The calibration schedule for an instrument should be kept at the instrument to show how long ago the instrument was calibrated. It is recommended to keep control charts on auditing tools. Tool control charts allow current

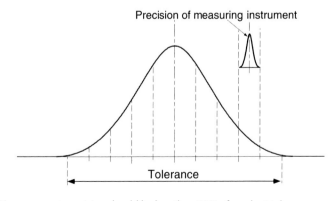

Figure 6.6 Measurement precision should be less than 10% of product tolerance

process data to be accurately compared to historical data and allow the operator to correct data for possible bias. Tool control charts also provide an indication of the need for calibration or repair.

Operator error can be reduced by training operators to use uniform and consistent measuring procedures. These procedures should be documented and provide enough detail to prevent different measuring techniques. Operator error can often be largely eliminated by automating the measurement. Instruments that automatically record their measurements will eliminate transposition and transcription errors. An additional advantage of automated measurements is speed, which means that more measurements can be made in the same period of time. Therefore, it is recommended to use automated measurements as much as possible.

6.4 Quantifying Measurement Variation

All measurements are subject to variability. For accurate process analysis it is necessary to determine whether the variability due to measurement is only a small fraction of the observed variability. Detailed procedures have been developed to assess the measurement error quantitatively. For a more complete discussion on the subject the reader is referred to the *Measurement System Analysis Manual* [17]. The degree of precision in a measurement process is generally designated as the standard deviation of measurement error. Common cause variation contributing to the precision is a combination of variation due to repeatability and reproducibility.

Repeatability is the variation obtained when one person measures the same quantity several times using the same measuring instrument; see Fig. 6.7.

Reproducibility is the variation in measurement averages due to differences among instruments, operators, and so on; see Fig. 6.8.

The standard deviation of the repeatability can be estimated from the average range. In control chart applications the relationship is:

$$\hat{\sigma}_{rpt} = \overline{R}/d_2 \qquad (6.2)$$

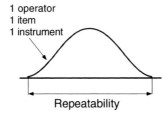

Figure 6.7 Illustration of repeatability

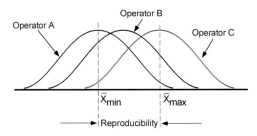

Figure 6.8 Illustration of reproducibility

The reciprocal values of constant d_2 are given in Table 6.1.

Table 6.1 Factors $1/d_2$ for Converting the Average Range, $\bar{R}$, into Estimated Standard Deviation $\hat{\sigma}_x$

	n = 1	2	3	4	5	8	10	>10
k = 2	.709	.781	.813	.826	.840	.855	.862	.885
3	.524	.552	.565	.571	.575	.581	.581	.592
4	.446	.465	.472	.474	.476	.481	.481	.485
5	.403	.417	.420	.422	.424	.426	.427	.429
6	.375	.385	.388	.389	.391	.392	.392	.395
7	.353	.361	.364	.365	.366	.368	.368	.370
8	.338	.344	.346	.347	.348	.348	.350	.351
9	.325	.331	.332	.333	.334	.334	.336	.337
10	.314	.319	.322	.323	.323	.324	.324	.325

n is the number of parts measured (number of samples)
k is the number of times each part was measured (sample size)
Courtesy of [3]

The estimated standard deviation of the reproducibility is determined from:

$$\hat{\sigma}_{rpd} = (\bar{x}_{max} - \bar{x}_{min})/d_2 \tag{6.3}$$

Constant d_2 is determined by the number of parts measured (n) and the number of times each part was measured; Table 6.1 shows values of $1/d_2$ for subgroup sizes from one to ten and up. The standard deviation of the measurement error is obtained from the relationship:

$$\sigma_e = \sqrt{\sigma_{rpt}^2 + \sigma_{rpd}^2} \tag{6.4}$$

The total measurement variation is a combination of variation due to precision, accuracy, and stability. This is illustrated in Fig. 6.9.

In general, a measurement process is considered capable when it is stable and its measurement error does not account for a significant portion of the tolerance band. The short-term precision should be less than 10% of the tolerance:

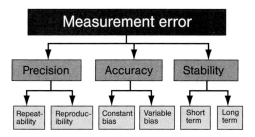

Figure 6.9 Components of measurement error

$$\frac{6\sigma_e}{USL - LSL} < 0.10 \qquad (6.5)$$

where σ_e is the short-term precision.

Short term is typically expressed in hours, while long term is expressed in days, weeks, or even months. The long-term precision should be less than 25% of the tolerance:

$$\frac{6\sigma_{el}}{USL - LSL} < 0.25 \qquad (6.6)$$

where σ_{el} is the long-term precision.

A number of different procedures have been developed over the years to analyze measurement systems. A collection of the Institute of Standards papers [4] describes procedures developed at this organization. Another useful reference is the Western Electric Statistical Quality Control Handbook [5]. The procedure for analyzing measurement systems described in this handbook is often referred to as the "Western Electric" method. Another good reference is the *General Motors Statistical Process Control Manual*. In the following, we will discuss the methods described in the GM manual in more detail.

6.4.1 Short Method for Gage R&R

Gage R&R refers to repeatability and reproducibility. This method requires a set of five parts to be measured by two operators. Each part is measured only once. A drawback of the short method is that repeatability cannot be isolated from reproducibility. Thus, the results reflect a combination of both types of measurement error.

The parts are selected at random and should have different values. The results have to be recorded on a table as shown in Table 6.2. The difference between measurements of operators A and B is entered in the column labeled "Range (A–B)." Only positive numbers are entered. The average range, $\overline{R}$, is the sum of all ranges divided by the number of parts. In the example in Table 6.2, the average range $\overline{R} = 1.4$.

The gage error is calculated by multiplying the average range by a constant 4.33. With $\overline{R} = 1.4$, the result is GR&R = 6.1, where GR&R stands for gage repeatability and reproducibility

Table 6.2 Data Collection Sheet

Parts	Operator A	Operator B	Range (A−B)
1	4	2	2
2	3	4	1
3	6	7	1
4	5	7	2
5	9	8	1
		Sum of ranges	7

Average range $\bar{R} = \Sigma R/5 = 7/5 = 1.4$
Gage error (GR&R) $= 4.33 \times \bar{R} = 4.33 \times 1.4 = 6.1$

and represents the combined gage error. To convert this number to a percentage of tolerance, divide by the tolerance and multiply by 100. If the tolerance for the example above is 20, then the GR&R(%) becomes:

$$\text{GR\&R}(\% \text{ tolerance}) = (6.1 \times 100/20) = 30.5\%$$

This level of gage error is generally considered to be too high. If the process variability is known, the GR&R can be compared to the process variation rather than the tolerance. For example, if the difference between the upper and lower control limit is 12, then GR&R as a percentage of the 6σ process variation becomes:

$$\text{GR\&R}(\% \ 6\sigma_{process}) = (6.1 \times 100/12) = 50.8\%$$

Again, this level of gage error is considered too high in most cases.

6.4.2 Long Method for Gage R&R

This method allows separate determination of gage repeatability and reproducibility. The results of the study can also provide information on the causes of gage error. If lack of reproducibility is large compared to repeatability, possible causes can be:

- The operator may not be properly trained in how to use and read the gage instrument.
- The scale and the numbers on the gage dial may not be clear.

If lack of repeatability is large compared to reproducibility, the reasons may be:

- The gage instrument needs maintenance.
- A better gage may be required.
- The location of the part in the gage may not be optimum.

The number of operators, the number of trials, and the number of parts may be varied. However, the study should be conducted according to the following steps:

1. Refer to operators A, B, C, and so on. The parts should be numbered so that the operator cannot identify the number. This increases the chance that all parts are measured the same way. A typical number of parts is ten.
2. Calibrate the gage.
3. Let operator A measure all parts in random order and enter the results in column 1 of data collection sheet; see Table 6.3.
4. Repeat step 3 with operator B, C, and so on.
5. Repeat steps 3 and 4 for the number of trials required.
6. Steps 3 through 5 may be modified for large size parts, unavailability of parts, or when operators are on different shifts.
7. Enter the observations on a data collection sheet as shown in Table 6.3. Using these numbers, calculate gage R&R using the formulas on the gage R&R report such as shown in Table 6.4.

The average range is determined as follows:

$$\overline{R} = (\overline{R}_a + \overline{R}_b + \overline{R}_c)/3 = (0.06 + 0.07 + 0.06)/3 = 0.0633$$

With constant $D_4 = 2.574$ (see Table 7.2), the upper control limit for the range is:

$$UCL_R = \overline{R} \times D_4 = 0.0633 \times 2.574 = 0.163$$

The difference between the maximum $\overline{x}$ and minimum $\overline{x}$ is:

$$\Delta \overline{x} = \overline{x}_{max} - \overline{x}_{min} = 0.8350 - 0.7533 = 0.0817$$

Notes:
_____
_____
_____

Table 6.3 Gage Repeatability and Reproducibility Data Collection Sheet

Name and Part Number:	Gasket, GA-104
Characteristic:	PTFE
Specification:	0.6–1.0 mm
Gage Name:	Thickness Gage
Gage number:	X-2034
Gage Type:	0-10 mm
Date (month-day-year):	6-15-92
Performed by:	M. Schneringer

(continued)

6 Measurement

Table 6.3 Gage Repeatability and Reproducibility Data Collection Sheet *(continued)*

	1	2	3	4	5	6	7	8	9	10	11	12
	Operator A				Operator B				Operator C			
Sample	Trial 1	Trial 2	Trial 3	Range	Trial 1	Trial 2	Trial 3	Range	Trial 1	Trial 2	Trial 3	Range
1	0.55	0.60	0.55	0.05	0.50	0.55	0.55	0.05	0.55	.050	0.55	0.05
2	0.95	1.00	0.95	00.5	1.00	0.95	1.05	0.10	1.05	1.00	0.95	0.10
3	0.90	0.90	0.85	0.05	0.75	0.75	0.75	0.0	0.75	0.85	0.80	0.10
4	0.90	0.90	0.90	0.0	0.70	0.75	0.80	0.10	0.75	0.80	0.80	0.05
5	0.60	0.55	0.55	0.05	0.40	0.40	0.45	0.05	0.50	0.55	0.55	0.05
6	0.95	1.05	0.95	0.10	0.95	1.05	1.00	0.10	1.05	1.05	1.00	0.05
7	1.00	1.00	0.95	0.05	0.90	0.90	0.95	0.05	0.95	0.95	1.00	0.05
8	0.90	0.80	0.90	0.10	0.80	0.70	0.75	0.10	0.80	0.85	0.80	0.05
9	0.95	1.00	0.95	0.05	0.90	0.90	0.95	0.05	1.00	1.00	1.05	0.05
10	0.55	0.65	0.60	0.10	0.45	0.50	0.55	0.10	0.90	0.85	.85	0.05
Total	8.25	8.45	8.15	0.60	7.35	7.45	7.80	0.70	8.30	8.40	8.35	0.60
	→	8.25		$\overline{R}_a =$	→	7.35		$\overline{R}_b =$	→	8.30		$\overline{R}_c =$
		8.15	←	0.06		7.80	←	0.07		8.35	←	0.06
	sum = 24.85				sum = 22.60				sum = 22.05			
	$\overline{x}_a = 0.8283$				$\overline{x}_b = 0.7533$				$\overline{x}_c = 0.8350$			

Table 6.4 Gage Repeatability and Reproducibility Report

Measurement Unit Analysis
Repeatability—Equipment *(EV)*
$EV = \overline{R} \times K_1 = 0.0633 \times 3.05 = 0.193$
K_1 (two trials) = 4.56 and K_1 (three trials) = 3.05
%EV = 100(EV)/tolerance = 100 × 0.193/0.40 = 48.25%
Reproducibility—Appraiser Variation *(AV)*

$$\sqrt{(\Delta \overline{x} K_2 - (EV)^2/(nr)(nr)} = \sqrt{(0.0817 \times 2.70)^2 - 0.193^2/(10 \times 3)} = 0.2055$$

where n is the number of parts, $n = 10$, and r is the number of trials, $r = 3$; K_2 (two trials) = 3.65 and K_2 (three trials) = 2.70. When the value under the square root sign is negative, the appraiser variation defaults to zero, AV = 0
Thus $\overline{R} = 0.0633$ and $\Delta \overline{x} = 0.0817$.

% AV = 100(AV)/tolerance = 100 × 0.2055/0.40 = 51.38%

Repeatability and Reproducibility (R&R)

$$R = \sqrt{(EV)^2 + (AV)^2} = \sqrt{0.193^2 + 0.2055^2} = 0.282$$

Note: All calculations based 5.15σ (99% of the area under the normal curve)

6.4.3 Gage Accuracy

Accuracy is generally not as much of a problem as precision. To determine gage accuracy, the true measurement of the sample parts is needed. This can often be done with tool room or layout inspection equipment. A true average is determined from these readings. This is compared later to the operators observed averages ($\bar{x}_A, \bar{x}_B, \bar{x}_C$).

If the sample parts cannot be measured this way, the following method can be used: Measure one of the sample parts precisely on tool room or layout inspection equipment. Have one operator measure the same part at least ten times, using the gage being evaluated. The difference between the true measurement and the observed averages represents gage accuracy. To convert accuracy to a percentage of tolerance, multiply by 100 and divide by the tolerance.

If there is significant lack of accuracy, possible causes can be:

- There is an error in the master.
- The gage is worn.
- Calibration is improper.
- Gage is being used improperly by operator.
- Gage is made to the wrong dimension.
- Gage is measuring the wrong characteristic.

6.4.4 Gage Stability

Gage stability should be determined although it usually is not as much of a factor as repeatability and reproducibility. How gage stability is determined depends on how often the gage is used between normal calibrations. The need for calibration depends on factors such as time, the number of measurements taken, air pressure change, warm-up, among others. Given that these factors are known, the calibration frequency can be established to minimize gage error due to stability.

If a gage is used intermittently, then its stability can be determined at the same time as the gage R&R study is made. The gage must be calibrated before and after each trial to determine the amount of calibration change. This amount is the gage stability for that trial only. To determine the overall gage stability, the calibration change for each trial must be added up and divided by the number of trials:

gage stability = (sum of calibration changes)/(number of trials)

If the gage is normally used for relatively long periods of time without calibration, the stability can be determined without recalibration for each trial. In this case, another gage R&R study is conducted just before the time calibration is due. Gage stability is then the absolute value of the difference between the grand averages of all the measurements in the first and second study.

To convert gage stability to a percentage of tolerance, multiply by 100 and divide by the tolerance. If gage stability error is large, check for these possible causes:

- The gage is not calibrated as often as necessary.
- Air pressure regulator or filter may be needed for air gauging.
- Warm-up may be required for electronic gages.

6.4.5 Gage Linearity

Gage linearity is assessed by determining the gage accuracy through the expected operating range. At least two accuracy studies should be conducted, one at each end of the operating range. The gage linearity is:

gage linearity = largest accuracy value − smallest accuracy value

If the gage linearity error is large, check for the following possible causes:

- Gage not calibrated properly.
- Error in the minimum or maximum master.
- Gage is worn.
- Gage may have inherently nonlinear characteristics.

6.4.5.1 Guidelines

The criteria for acceptance of gage repeatability and reproducibility are:

GR&R (% tolerance) < 10% — Acceptable

GR&R (% tolerance) = 10% to 30% — Possibly acceptable, depending on importance of application, cost of the gage, cost of repairs, and so on.

GR&R (% tolerance) > 30% — Generally not acceptable

6.5 Graphical Method for Measurement Analysis

The methods for evaluating the measurement system outlined in the previous sections consist of tabulating a number of measurements by different operators and performing calculations on these data to determine repeatability and reproducibility. The tabular method of evaluating gage R&R is not always the most efficient, as discussed by Wheeler [35]. A more efficient way of evaluating gage R&R can be the control chart.

6.5.1 Traditional Gage R&R Method

In the following example three different operators measure the thickness of five parts; each part is measured twice. The results are presented in Table 6.5. The grand average of all the measurements is 75.8 units and the average range for the fifteen subgroups is 4.267 units. The average for operator A is 81.0, for operator B 72.5 and for operator C 73.9 units. The part averages are respectively 58.0, 106.167, 82.0, 84.833, and 48.0. The specification for the thickness of the parts is $T = \pm 20$ units; thus the total tolerance spread is forty units.

Repeatability is a measure of the variation in repeated measurements, in this case between the first and second measurement. The standard deviation of the repeatability variation can be determined by dividing the average range by the appropriate d_2 value:

$$\sigma_{rpt} = \frac{\bar{R}}{d_2} = \frac{4.267}{1.128} = 3.783 \text{ units} \tag{6.7}$$

This value traditionally has been multiplied with a factor 5.15. Using this approach, the repeatability is $5.15 \times 3.783 = 19.48$ units. This value is then divided by the specified tolerance and expressed as a percentage:

$$\frac{5.15\sigma_{rpt}}{\text{tolerance}} 100 = \frac{5.15 \times 3.783 \times 100}{40} = 48.7\% \tag{6.8}$$

This indicates that variation due to repeatability takes up 48% of the specified tolerance. This amount is much larger than would be desired.

Table 6.5 Measurement Results with Three Operators and Five Parts

Operator A						Average
Part number	1	2	3	4	5	
First reading	67	110	87	89	56	
Second reading	62	113	83	96	47	
Average operator A	64.5	111.5	85	92.5	51.5	81.0
Range operator A	5	3	4	7	9	5.6
Operator B						
First reading	55	106	82	84	43	
Second reading	57	99	79	78	42	
Average operator B	56	102.5	80.5	81	42.5	72.5
Range operator B	2	7	3	6	1	3.8
Operator C						
First reading	52	106	80	80	46	
Second reading	55	103	81	82	54	
Average operator C	53.5	104.5	80.5	81	50	73.9
Average A, B, and C	58.0	106.17	82.0	84.83	48.0	75.8
Range operator C	3	3	1	2	8	3.4

Reproducibility is a measure of the variation due to differences between operators. The standard deviation of the variation due to reproducibility can be determined by taking the range of the operator averages, R_0, and dividing by the appropriate d_2 value. The d_2 value is determined only by the number of data, n, in the group, in this example $n = 3$. The values for d_2 will be given in Table 7.2 of Chapter 7 (Control Charts); for $n = 3$, the value of $d_2 = 1.693$. The standard deviation of the reproducibility variation thus becomes:

$$\sigma_{rpd} = \frac{R_0}{d_2} = \frac{8.50}{1.693} = 5.021 \text{ units} \tag{6.9}$$

The standard deviation due to repeatability and reproducibility is determined by adding the squared values and then taking the square root of the sum:

$$\sigma_{R\&R} = \sqrt{(\sigma_{rpt}^2 + \sigma_{rpd}^2)} = \sqrt{3.783^2 + 5.021^2} = 6.287 \text{ units} \tag{6.10}$$

This R&R is traditionally multiplied by 5.15, divided by the specified tolerance, and expressed as a percentage. For this example we get:

$$R\&R(\%\text{tolerance}) = \frac{5.15 \sigma_{R\&R}}{\text{tolerance}} 100 = \frac{5.15 \times 6.287 \times 100}{40} = 80.95\% \tag{6.11}$$

Thus the combined R&R takes up 80.95% of the specified tolerance — this is unacceptably high.

6.5.2 Graphical Approach to Gage R&R

The results from the measurements can be plotted in an average and range chart; this will make the results much easier to interpret. Control charts will be discussed in more detail in Chapter 7. The data from Table 6.5 are shown in Fig. 6.10. The solid line on the range chart at value 4.27 is the central line of the range chart; this is the average range, $\bar{R}$. The upper control limit or UCL of the range chart is determined by multiplying the average range with a factor D_4. Thus UCL $= \bar{R} \times D_4$; the control limits are represented by dashed lines. The value of factor D_4 depends only on the number of data in the subgroup; in this case $n = 2$ and $D_4 = 3.267$. Thus UCL of the range chart becomes UCL $= 4.267 \times 3.267 = 13.9$. The lower control limit of the range chart or LCL is determined by LCL $= \bar{R} \times D_3$. Like D_4, factor D_3 only depends on the number of data in the subgroup; in this case $D_3 = 0$ and, thus the LCL $= 0$. The values of D_3 and D_4 can be taken from Table 7.2 of Chapter 7, Control Charts.

The solid line in the average chart is the grand average ($\bar{\bar{x}} = 75.8$). The upper control limit is determined from UCL $= \bar{\bar{x}} + A_2 \bar{R}$, where A_2 depends only on the number of data in the subgroup and is given in Table 7.2.

In this example, $A_2 = 1.88$ and UCL $= 75.8 + 1.88 \times 4.267 = 83.82$. The lower control limit or LCL is determined from:

$$\text{LCL} = \bar{\bar{x}} - A_2 \bar{R} = 75.8 - 1.88 \times 4.267 = 67.78.$$

Both control limits are represented by dashed lines.

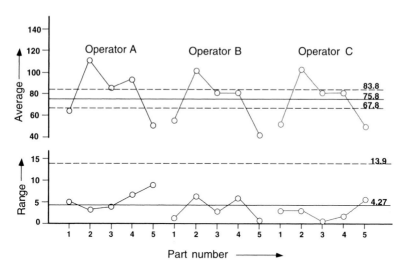

Figure 6.10 Average and range chart for gage R&R study

The range chart shows no signs of special problems; it appears to be in control. If the range chart is out of control, then there is a problem with the measurement process that needs to be analyzed and corrective action needs to be taken. When the range chart is in control, the average range may be divided by d_2 to determine repeatability; this is also called the test-retest error or replication error.

The average chart indicates how well the instrument performs in the measurements made in the study. The running record of the average chart creates a horizontal band that represents the product variation. The control limits show how much of the product variation is obscured by the measurement error. It is obvious that the product variation is much larger than the measurement error; the average chart appears to be out of control! In this application, however, it is desirable to have the product variation larger than the measurement error. The smaller the measurement error relative to the product variation, the more useful the measuring instrument! The usefulness of the instrument can be quantified by a factor called the discrimination ratio, D_r.

The discrimination ratio is a measure of how wide the running record is relative to the control limits of the average chart. The running record represents the product variation, while the control limits are determined only by the repeatability. The discrimination ratio defines the number of nonoverlapping categories within the natural process limits that the product could be sorted into using these measurements. For example, if the discrimination ratio is six, then the product can be sorted into six nonoverlapping categories. The expression that can be used to determine the discrimination ratio will be given later, see Eq. 6.16.

From the averages chart it can be seen that the measurements from operator A are higher than those of operators B and C. The running record for the three operators shows good parallelness, indicating that there is no significant interaction between operator and part. This illustrates how useful an average and range chart can be in analyzing data from a R&R study. It can be used to:

- Check for consistency of repeatability
- Provide a simple estimate of repeatability
- Show the utility of the instrument to measure a specific product
- Uncover any potential differences between operators
- Check for interactions between operators and parts

6.5.3 Differences Between Operators

The value of reproducibility derived earlier does not give an indication of the actual differences between operators. We can use a control chart to assess the actual differences between operators. The repeatability value determined earlier, 3.783 units, characterizes the measurement error associated with a single value. The operator averages from Table 6.5 are based on ten readings each. The estimated standard deviation of the repeatability of the operator averages can be determined from:

$$\sigma_{rpt\bar{x}} = \frac{\bar{R}}{d_2\sqrt{n_r}} = \frac{4.267}{1.128\sqrt{10}} = 1.196 \qquad (6.12)$$

where d_2 for a subgroup of two is $d_2 = 1.128$ as determined from Table 7.2 and n_r is the number of readings, $n_r = 10$.

With this information we can determine the limits for the operator averages. With the grand average 75.8, the upper limit becomes UCL = 75.8 + 3 × 1.196 = 79.39 and the lower limit LCL = 75.8 − 3 × 1.196 = 72.21. We can now chart the averages of the three operators, as shown in Fig. 6.11.

It is quite clear in Fig. 6.11 that operator A is above the upper limit, while the other two operators are within the limits. This indicates that operator A does measure the parts differently from operators B and C. Based on these results it would make sense to look closely at how operator A makes the measurements; it is quite possible that operator A is not using correct techniques and requires additional training.

6.5.3.1 Measurement Error and Product Variation

The proportion of the variance in the product measurements attributable to product variation can be quantified by the intraclass correlation coefficient (ICC). When the ICC is subtracted from 1.0, the result will express the proportion of the measurement variation attributable to measurement error. The ICC is determined from σ_m^2, the variance of product measurements, and σ_e^2, the variance in the measurement process. An example of determining ICC will be given based on the numbers in Table 6.5.

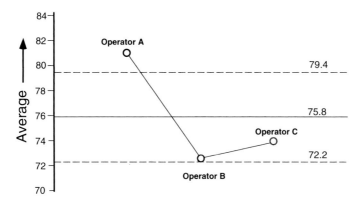

Figure 6.11 Control chart for operator averages

The part variation is determined by taking the range of the part averages, R_p, and dividing by the appropriate value of d_2 (see Table 7.2 of Chapter 7, Control Charts). For five parts, the appropriate value of $d_2 = 2.326$. The part variation thus becomes:

$$\text{part variation } \frac{R_p}{d_2} = \frac{58.167}{2.326} = 25.01 \tag{6.13}$$

Part variation can be used as a measure of σ_m, the standard deviation of product measurements. Thus, the value of $\sigma_m = 25.01$ and $\sigma_m^2 = 625.5$. The estimated standard deviation of the measurements due to repeatability was earlier determined to be 3.783 units. If we assume for the moment that the only measurement variation is due to repeatability, then the estimated variance of the measurement process is:

$$\sigma_e^2 = 3.78^2 = 14.311 \tag{6.14}$$

The ICC, represented by character r_i, can be determined from the following equation:

$$r_i = 1 - \frac{\sigma_e^2}{\sigma_m^2} = 1 - \frac{14.31}{625.5} = 0.9771 \tag{6.15}$$

This means that almost 98% of the variation in these measurements is due to product variation and only 2% is due to repeatability. The discrimination ratio, discussed earlier, can be determined from the intraclass correlation coefficient:

$$D_r = \sqrt{\frac{1+r_i}{1-r_i}} = \sqrt{\frac{\sigma_e^2}{\sigma_e^2}} \tag{6.16}$$

A ICC value of $r_i = 0.9771$ results in a $D_r = 9.3$. This means that one can use nine product categories within the natural process limits. If we incorporate both the repeatability and reproducibility, we can use the $\sigma_{R\&R}$ value determined earlier, see Eq. 6.10. Thus the measurement variance becomes:

$$\sigma_e^2 = 6.287^2 = 39.52 \qquad (6.17)$$

The intraclass correlation coefficient can now be determined from:

$$r_i = 1 - \frac{\sigma_e^2}{\sigma_m^2 + \sigma_{rpd}^2} = 1 - \frac{39.52}{650.71} = 0.9393 \qquad (6.18)$$

This means that 94% of the variation in the measurements is due to product variation and 6% is due to repeatability and reproducibility. With $r_i = 0.9393$, the corresponding discrimination ratio is $D_r = 5.7$. With this value of the discrimination ratio one can use six product categories within the natural product limits.

6.6 Measurements in Injection Molding

The degree of process control that can be attained in the injection molding process is strongly determined by how well measurements are made. This means not only measuring the appropriate process parameters, but also using the right sensors and transducers and placing these in the correct locations.

6.6.1 Important Process Parameters

In a typical injection molding process, there are around twenty process parameters that should be monitored to control the process and ensure consistency. The most important parameters, more or less in order of priority, are:

Product parameters:
- Product dimensions
- Product weight
- Product appearance

Main SPC process parameters:
- Cavity pressure
- Injection pressure
- Fill time

Additional process parameters:
- Melt temperature
- Mold temperature
- Injection energy

- Position at start of injection
- Position at transition to hold pressure
- Cooling time
- Recovery time
- Total cycle time
- Extruder barrel temperatures
- Barrel heaters power consumption
- Screw rotational speed
- Screw power consumption during rotation

The preferred method is to measure on-line on a continuous basis, that is, to use real-time monitoring. If the measurement data are fed to a computer, then ranges, control limits, standard deviations, and so on, can be calculated almost instantaneously. Thus control charts can be generated and displayed within seconds of the actual measurement. This provides real-time SPC, which can be very useful in accurate monitoring and control of the process.

6.6.1.1 Product Characteristics

Various measurement techniques are available to determine product [6]. These are based on laser, optical, electromagnetic, calipers and micrometers, ultrasound, inductive, and pneumatic-inductive measurements. For complicated parts, special fixtures with gages can be developed that measure various dimensions simultaneously or in rapid succession. Such devices can save much manual measurement time, reduce error, and allow automation.

Wall thickness is often measured with ultrasound gages. The ultrasound measurement is basically a measurement of time; it measures how long a sound wave takes to travel from the outside of the wall to the inside and back. It must be realized that the propagation of sound waves not only depends on the type of material, but also on the material temperature. As a result, considerable measurement error can occur when the stock temperature of the material changes.

The choice of the measurement method will depend on several factors, such as the shape and dimensions of the product, the tolerance, the cost of gage, user friendliness, and so on. Measurement of the product dimensions is often the most critical measurement in the process. As a result, the gage should be subjected to a careful R&R study to make sure that the gage is acceptable for the application.

6.6.1.2 Pressure

Pressure is probably the most important process variable in injection molding. In most machines, both the hydraulic pressure and the plastic melt pressure are important.

Obviously, on all electric injection molding machines only the plastic melt pressure is at issue. The plastic melt pressure will vary in different parts of the injection molding machine. For instance, the melt pressure at the screw tip will be higher than the melt pressure in the cavity. This is due to the pressure losses in the runner system. A typical pressure-time profile during an injection molding cycle is shown in Fig. 6.12.

Obviously, the pressure changes dramatically over the course of a cycle. The pressure rises rapidly during injection. Excessive pressure buildup in the plasticating unit can be avoided with a pressure relief device, such as a rupture disk. The screw tip pressure usually is constant during the hold time, while the cavity pressure decreases during hold time. After the material in the gates freezes, the screw tip pressure can be reduced to the backpressure required during screw recovery. The pressure values often used in control charts are the peak cavity pressure and injection pressure. In Fig. 6.12, the injection pressure is equal to the hold pressure; it is possible for the injection pressure to be higher than the hold pressure.

Most pressure transducers used in plastic processing are electronic transducers that use strain gages on a diaphragm to measure pressure. The strain gage diaphragm is placed a considerable distance away from the diaphragm that is actually in contact with the polymer melt. This is done to minimize the effect of the high temperature of the polymer melt. Two types of electronic strain gage transducers are used: the capillary type and the pushrod type; see Fig. 6.13.

In the capillary type a fluid is used to hydraulically couple the melt diaphragm and the strain gage diaphragm. The fluid is usually mercury. A drawback of these electronic transducers is that they are susceptible to damage. In mercury filled transducers this will release mercury in the polymer melt and the work place which is highly undesirable. The damage is due to the thin diaphragm in contact with the melt. A pressure transducer that is

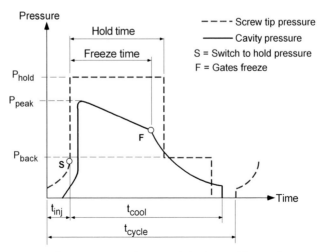

Figure 6.12 Screw tip and cavity pressure during an injection molding cycle

not removed before a barrel is cleaned, is likely to be damaged. Since these transducers are not inexpensive, typically around $1000, frequent damage can be costly.

On the other hand, there are piezoresistive pressure transducers that are more rugged and less susceptible to damage, due to a thicker diaphragm. These newer types of pressure transducers [7] feature silicon-on-sapphire technology to obtain good accuracy, stability, and sensitivity over a wide temperature range up to 370 °C (700 °F).

Another type of pressure transducer is the optical pressure transducer. This transducer is based on optical intensity modulation and was developed by the German company FOS Meßtechnik. Light from an infrared LED (light emitting diode) is coupled to a fiber optic light guide. The light is guided to the sensor head and the reflected light intensity is compared to the incoming light intensity. As the diaphragm of the transducer is deformed by pressure, the ratio of reflected and incoming light intensity changes and, thus the pressure is measured. The advantages of this type of transducer are its rugged design, good dynamic response, high temperature capability, and small measurement error. The US company Dynisco also developed an optical pressure transducer that was commercially available for some time, but was taken off the market in 1996. A comparison of various pressure transducers is shown in Table 6.6.

Electronic pressure transducers are sometimes equipped with a temperature sensor. These transducers are typically referred to as melt pressure/temperature (P/T) transducers. However, it should be realized that in capillary transducers the temperature sensor is usually placed a considerable distance from the diaphragm.

Therefore, the temperature sensed is not a good indication of the actual melt temperature. There can be considerable differences between the sensed temperature and the actual melt temperature, as much as 50 °C (90 °F) or more [8].

Piezoelectric transducers are not suitable to measure static pressure; however, they are quite useful for dynamic pressure measurement. As a result, they are commonly used in injection molding. They are limited to temperatures of 120 °C (248 °F); it is important, therefore, to

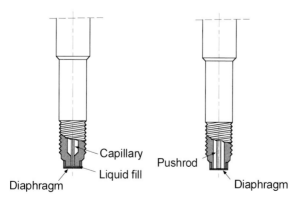

Figure 6.13 Capillary transducer (left) and pushrod transducer (right)

make sure that the transducer temperature stays below that level when piezoelectric transducers are used.

Table 6.6 Comparison of Various Pressure Transducers

Pressure measurement	Reliability/ robustness	Temperature range (°C)	Temperature sensitivity	Dynamic response	Total error %
Bourdon	Poor	200	Poor	Poor	3–5
Pneumatic	Good	400	Poor	Poor	1.5
Capillary* strain gage	Fair	400	Fair	Fair	0.5–3
Pushrod strain gage	Fair	400	Poor	Fair	3
Piezoelectric**	Good	120	Fair	Excellent	0.5–1.5
Piezoresistive	Good	400	Good	Good	0.2–0.5
Optical	Good	600	Good	Good	1

* Environmental concerns with mercury filled transducers
** Limited to dynamic pressure measurement

6.6.1.3 Temperature

The melt temperature is the most important temperature in the injection molding process. It is usually measured at the end of the extruder with an immersion temperature sensor. For a good melt temperature measurement it is important that there be good contact between the polymer melt and the sensor. That is why pressure/temperature transducers are not suitable for melt temperature measurement, as discussed earlier. Also, flush mounted melt temperature sensors are not too useful, because the measured temperature is mostly representative of the metal wall temperature. Various types of melt temperature probes are shown in Fig. 6.14.

Barrel and mold temperatures are also quite important in the injection molding process. These are usually measured with a bayonet-type thermocouple (TC), as shown in Fig. 6.15. Tests [10] with conventional thermocouples have shown that significant measurement errors can be caused by:

- Air currents around the plasticating unit can cause errors of up to 50 °C (90 °F).
- Insufficient depth of the TC well can cause errors of up to 10 °C (18 °F). The depth should be at least 25 mm (1 in).

Special thermocouples have been designed to minimize conduction error. These TCs typically have an exposed sensing element on a ceramic substrate for thermal insulation.

When measuring barrel temperatures, it is important to measure close to the inside diameter of the barrel. After all, it is the polymer temperature that is really important in the molding process. It should be realized, however, that in some commercial machines the

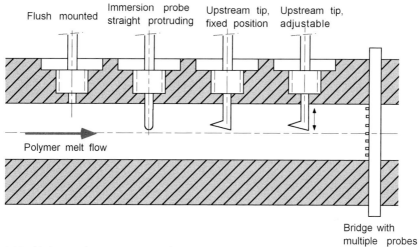

Figure 6.14 Various melt temperature probes

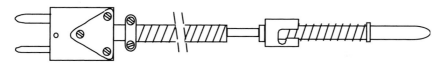

Figure 6.15 Bayonet-type thermocouple

temperature sensor is placed close to the outside diameter of the barrel. In fact, in some instances the temperature sensors are placed in the barrel heaters. This will yield good temperature control; however, the *wrong* temperature is being controlled. We should be concerned about the polymer temperature and therefore, we should measure and control the temperature as close to the polymer as reasonably possible.

Melt temperature can also be measured by infrared (IR) probes. The advantage of IR melt temperature measurement is that very rapid changes in melt temperature can be measured; a typical response time is about 10 milliseconds. A disadvantage of the IR melt temperature measurement is that it is more expensive than the immersion thermocouple.

In analyzing the process, one should know what type of sensors and transducers are used and where they are located. If this information is not available, entirely incorrect conclusions could be drawn. For instance, with TCs in the barrel heater, the temperature control would appear to be excellent. At the same time, however, the temperature at the inside barrel could fluctuate wildly, causing major upsets in the process. The best situation is to have both a deep and shallow barrel temperature measurement, as long as the deep well temperature is the one being controlled. This allows determination of the radial temperature gradient, and, thus of the amount of heat being added or removed from the process. There are also dual sensor temperature control systems that use both shallow and deep temperature sensors. The deep temperature is the one being controlled; however, the shallow temperature participates in the control process through a cascade loop (see [11]).

6.6.1.4 Screw Speed

The recovery rate of the extruder is determined directly by the screw speed. It is important, therefore, that the screw speed can be accurately measured and controlled. The screw speed should be measured and displayed to at least 0.1 rpm, preferably 0.01 rpm or even less. In other words, the sensitivity and resolution of the screw speed measurement should be 0.1 rpm or better. Assuming a full scale of 200 rpm, this represents 0.05 % FS.

6.7 Measurements in Extrusion

The degree of process control that can be attained in the extrusion process is strongly determined by how well measurements are made. This means not only measuring the appropriate process parameters, but also using the right sensors and transducers and placing these in the correct location.

6.7.1 Important Process Parameters

In a typical extrusion process, there are between 50 and 100 process parameters that should be monitored to control the process and ensure consistency. The most important parameters, more or less in order of priority, are:

1. Extrudate dimensions
2. Diehead pressure (after screen pack)
3. Barrel pressure (before screen pack)
4. Polymer melt temperature
5. Motor load
6. Screw speed
7. Take-up speed
8. Power consumption of various heating zones
9. Cooling rate at each cooling zone
10. Barrel and die temperatures

The two most important measurements are extruded product dimensions and diehead pressure. Obviously, product dimensions have to be monitored to make sure that the product is within dimensional specifications. The preferred method is to measure on-line on a continuous basis, that is, to use real-time monitoring. If the measurement data are fed to a computer, then averages, ranges, control limits, standard deviations, and so on, can be calculated almost instantaneously. Thus control charts can be generated and displayed within seconds of the actual measurement. This provides real-time SPC, which can be very useful in accurate monitoring and control of the process.

6.7.1.1 Extrudate Dimensions

Various measurement techniques are available to determine extruded product dimensions [6]. These are based on laser, optical, electromagnetic, ultrasound, inductive, and pneumatic-inductive measurements. The first three are used mostly to determine outside dimensions of extruded products; the last four are used to measure wall thickness. Laser gages are frequently used to measure the outside diameter of tubing, pipe, and coated wire. The resolution varies from 0.001 to 0.01 mm, depending on the range.

Wall thickness is often measured with ultrasound gages. This works well on pipe and large diameter tubing; however, on small (diameter less than 10 mm) thin walled tubing this measurement is problematic. The ultrasound measurement is basically a measurement of time; it measures how long a sound wave takes to travel from the outside of the wall to the inside and back. It must be realized that the propagation of sound waves not only depends on the type of material, but also on the material temperature. As a result, considerable measurement error can occur when the stock temperature of the material changes. This can happen when the line speed changes, when extruder temperatures are changed, when the cooling water temperature changes, and so on. The measuring head is typically submerged in water to provide good coupling between the head and the product to be measured. Care must be taken that clean water is used without air bubbles present. Air bubbles can cause considerable measurement error.

The choice of the measurement method will depend on several factors, such as the shape and dimensions of the extruded product, the tolerance, the cost of gage, userfriendliness, and so on. Measurement of the product dimensions is often the most critical measurement in the extrusion process. As a result, the gage should be subjected to a careful R&R study to make sure that the gage is acceptable for the application.

6.7.1.2 Pressure

The pressure measurement is important for two main reasons: process control and safety. The diehead pressure, that is, the pressure after the screen pack and the breaker plate, determines the output. A steady pressure will give a steady output; however, a varying pressure will result in a varying output. Output itself is not easily measured in the extrusion process. However, pressure can be measured rather easily.

It is most important to determine how diehead pressure changes with time; simple instantaneous pressure readout is not very meaningful. The variation of pressure with time can be captured on a simple chart recorder or, better, on a CRT when a computerized data acquisition system (DAS) is available.

The change in output, ΔQ, resulting from a change in pressure, ΔP, can be determined from the following relationship:

$$\Delta Q(\%) = \Delta P(\%)/n \tag{6.19}$$

In this expression n is the power law index of the material. For polymer melts n varies between 0 and 1; for most polymers n ranges from 0.3 to 0.7. This means that a 2% pressure fluctuation for a polymer with a power law index $n = 0.4$ will result in an output fluctuation of 5%! In other words, the output fluctuation is considerably larger than the pressure fluctuation. This is true for all polymers. Output fluctuations will tend to cause variations in the extruded product dimensions. Thus it is very important to keep pressure fluctuation as small as possible. If a circular product is extruded, the percent change in the diameter resulting from output variation will be about half the percent output variation. Thus an 5% output variation will cause approximately a 2.5% diameter variation.

As an example, let us consider the extrusion of a 10 mm diameter rod. The specifications are 10.3 to 9.7 mm; this is 10.0 mm ± 3%. The maximum output variation that can be tolerated in this case is about 6%; this is twice the percentage of the specifications. If the material is LDPE with a power law index $n = 1/3$, a 2% pressure fluctuation will cause a 6% output fluctuation. In this case, the maximum pressure fluctuation that we can tolerate is 2%. This, of course, assumes that there are no other sources of dimensional variations, which is a questionable assumption. Therefore, we would want to keep the pressure fluctuation to less than 2% in order to keep the extruded product within specifications. Clearly, if the actual pressure fluctuation is 5%, there is little or no chance that we can extrude the product within specifications.

The safety aspect of pressure measurement is critically important. Under certain conditions (e.g., cold start-up) an extruder can generate dangerously high pressures, much in excess of 70 MPa (10,000 psi). MPa stands for megapascal, which is 1,000,000 pascals. One pascal is one newton per square meter. If no pressure relief safety is present, such high pressures can force the die from the extruder at high speed. Any person in the path of this die is likely to be severely injured. Unfortunately, a number of serious accidents have happened as a result of excessive pressure buildup. All extruders should be equipped with a pressure relief device, such as a rupture disk or a shear pin. Unfortunately, this is not always the case. It is extremely important, therefore, that a properly working pressure measurement be available on every operating extruder, preferable with a high limit alarm and automatic shut down. The most critical location of the pressure transducer for determining excessive pressures is just before (upstream) the breaker plate. Thus for process control the pressure measured behind the breaker plate is most important, while for safety the pressure just before the breaker plate is most important. It is advisable, therefore, to have at least two pressure transducers on an extruder. The difference in the two pressure readings is a good measure of the buildup of contamination on the screen pack.

6.7.1.3 Screw and Take-Up Speed

The output of the extruder is determined directly by the screw speed. It is important, therefore, that the screw speed can be accurately measured and controlled. The screw speed should be measured and displayed to at least 0.1 rpm, preferably 0.01 rpm or even less. In other words, the sensitivity and resolution of the screw speed measurement should be 0.1 rpm or better. Assuming a full scale of 200 rpm, this represents 0.05% FS. This is particularly important when an extruder is operated at low screw speed. For example, if an extruder runs at 5 rpm and the resolution of the screw speed is 1 rpm, the actual screw speed could vary as much as 20% without any indication from the screw speed readout. With a resolution of 0.1 rpm, the actual screw speed could still vary as much as 2% without an indication from the screw speed readout.

Another important consideration is the speed regulation of the extruder drive and the take-up drive. Typical brush type DC drives with tachometer feedback have a speed regulation of 1% FS. Assuming a full speed of 100 rpm, this means that at 10 rpm the screw speed can vary ± 1 rpm. This means ± 10%, a total variation of 20%! This is unacceptable in most extrusion operations. Much better speed regulation can be obtained with digital drives, such as the brushless DC drive or the DC brush with pulse encoder. These drives have a speed regulation of 0.01% FS or better. That means that even at low screw speed, these drives can maintain a very steady screw speed.

Clearly, good speed regulation is equally important for the take-up. If the extruder output is steady but the take-up speed is varying, the extrudate dimensions will vary as well, because the drawdown of the extruded product will change as the take-up speed changes. It is recommended, therefore, that both the extruder and the take-up device be equipped with drives that provide a speed regulation of 0.01% FS or better. Since digital drives nowadays are no more expensive than the old-fashioned DC brush drives with or without tachometer feedback, there is little reason not to use digital drives.

7 Control Charts

7.1 Introduction

As discussed in Chapter 4, control charts are one of the principal tools in statistical process control. Variations in a process can occur as a result of common causes and assignable causes. Common cause variation is inherent to any process that is stable over time. Assignable cause variation results from significant and identifiable changes in the process, such as a different screw in the plasticating unit, a change in the polymer drying conditions, a new operator, a new lot of raw material, and so on. In most cases, reducing variation by eliminating assignable causes is more practical than by eliminating common causes. Process control charts are the best tool for identifying and eliminating assignable cause variation.

Control charts show how process data evolve over time and allow determination of natural process variability. Three important uses of control charts are:

- They can identify assignable causes when they occur.
- They can determine whether improvement action really reduces process variation.
- They can be used to determine the true process capability.

7.2 Control Charts for Variables Data

Control charts can be used for both variables and attributes data. Variables control charts are more sensitive to changes in the measured values and, therefore, are better for process control. The various control charts for variables are shown in Table 7.1.

x and Rm: The individual measurement, x, and moving range, Rm, chart is used when process control is based on individual readings. This can happen when the measurements are expensive. Sometimes the moving range chart is combined with a moving average chart.

Table 7.1 Control Charts for Variables Data

Number of data in subgroup	Type of control charts
$n = 1$	x and Rm
$1 < n < 7$	$\bar{x}$ and R, $\tilde{x}$ and R, $\tilde{x}$ and x, $\bar{x}$ and s, CuSum
$n > 7$	$\bar{x}$ and s, CuSum

$\bar{x}$ and R: The mean, $\bar{x}$ and range, R, chart is the most common control chart. It is discussed below in more detail.

$\tilde{x}$ and R: The median, $\tilde{x}$ and range, R, chart is similar to the $\bar{x}$ and R chart. The median is the middle value when data are arranged according to size. Sometimes the median chart is used by itself without the range chart.

$\tilde{x}$ and x: The median, $\tilde{x}$, and individual measurement, x, chart is a special chart used for family processes. This is especially useful for injection molding operations with multicavity molds. It allows separation of global and local factors; this is discussed in more detail in Section 8.5 in the section on special SPC techniques in injection molding.

$\bar{x}$ and s: The mean, $\bar{x}$, and standard deviation, s, chart is a more accurate indication of process variability than the $\bar{x}$ and R chart, particularly with a larger size subgroup. The standard deviation, s, is more difficult to calculate than the range, R; however, this is immaterial when the data are processed by a computer.

CuSum: A cumulative sum (CuSum) chart shows the cumulative sum of each $\bar{x}$ value minus the nominal value. This chart is more sensitive to a sustained shift from the nominal.

7.2.1 The $\bar{x}$ and R Chart

Since the $\bar{x}$ and R chart is so commonly used in SPC, we will treat this chart in depth. The methods of usage and analysis for other charts are similar to those for the $\bar{x}$ and R chart. Therefore what we learn about the analysis and interpretation of $\bar{x}$ and R charts can be applied to other charts as well.

The creation of an $\bar{x}$ and R chart requires several steps. We will go through this process step by step.

Step 1: Select the subgroup size and determine how often samples are taken. A common subgroup size is $n = 5$, with n the number of measurements in the subgroup.

Step 2: Record the raw data on the control chart. The data in each subgroup are listed in columns. If the number of subgroups is k, then the total number of measurements is $n \times k$. Proper determination of control limits requires at least 25 subgroups ($k \geq 25$).

Step 3: Calculate the sum Σx, the mean, $\bar{x}$, and the range, R, for each subgroup using the following equation:

$$\sum_{i=1}^{i=n} x_i = x_1 + x_2 + x_3 + \ldots + x_n \qquad (7.1)$$

$$\bar{x} = \sum x_i / n \qquad (7.2)$$

$$R = x_{max} - x_{min} \qquad (7.3)$$

$\sum x_i$ a simplified notation for $\sum_{i=1}^{i=n} x_i$ which is a shorthand notation for the sum of the x-values from x_1 all the way through x_n.

Step 4: Calculate the central lines for the $\bar{x}$ and R chart:

$$\bar{\bar{x}} = \frac{(\bar{x}_1 + \bar{x}_2 + \bar{x}_3 + \ldots + \bar{x}_k)}{k} \qquad (7.4)$$

$$\bar{R} = \frac{(R_1 + R_2 + R_3 + \ldots + R_k)}{k} \qquad (7.5)$$

where:

$\bar{R}$ = the average range

$\bar{\bar{x}}$ = the average of all samples or the "grand average"

Step 5: Calculate the control limits for the $\bar{x}$ and R chart. First the upper control limit for the $\bar{x}$ chart:

$$\text{UCL } \bar{x} = \bar{\bar{x}} + A_2 \bar{R} \qquad (7.6)$$

The lower control limit for the $\bar{x}$ chart is:

$$\text{LCL} = \bar{\bar{x}} - A_2 \bar{R} \qquad (7.7)$$

The upper control limit for the range chart is:

$$\text{UCL range} = D_4 \bar{R} \qquad (7.8)$$

The lower control limit for the range chart is:

$$\text{LCL range} = D_3 \bar{R} \qquad (7.9)$$

UCL is the upper control limit and LCL is the lower control limit. The constants A_2, d_2, D_4, and D_3 can be taken from Table 7.2; these factors only depend on the size of the subgroup. Factor D_3 is zero for $n = 2$ to $n = 6$; this means that the lower control limit for the range chart is zero for $n = 2$ to $n = 6$. Also, the central line of the range chart will not be centered between the control limits for subgroups smaller than $n = 7$. For subgroups of 7 and larger D_3 is nonzero, and thus the LCL is also nonzero and the central line will be centered between the control limits.

Step 6: Plot the sample averages and the sample ranges. Use the following guidelines:

- The grand average, $\bar{\bar{x}}$ should be near the center of the $\bar{x}$ chart.
- The UCL for the $\bar{x}$ chart should be at about 2/3 from the $\bar{\bar{x}}$ to the top of the scale.
- The UCL and LCL have to be equidistant from $\bar{\bar{x}}$.
- The scale of the R chart should extend from 0 to about 4/3 times the UCL range.
- The points should be plotted as solid dots.
- Connect the points for easier pattern identification.

Step 7: Draw the central lines and control limits on the charts. The central lines should be solid, the control limits dashed. The lines should be labeled with their numerical value.

Table 7.2 Factors for Determination of Control Limits

Subgroup size n	Factor A_2	Constant d_2	Factor D_3	Factor D_4
2	1.880	1.128	0	3.267
3	1.023	1.693	0	2.574
4	0.729	2.059	0	2.282
5	0.577	2.326	0	2.114
6	0.483	2.534	0	2.004
7	0.419	2.704	0.076	1.924
8	0.373	2.847	0.136	1.864
9	0.337	2.970	0.184	1.816
10	0.308	3.078	0.223	1.777
11	0.285	3.173	0.256	1.744
12	0.266	3.258	0.283	1.717
13	0.249	3.336	0.307	1.693
14	0.235	3.407	0.328	1.672
15	0.223	3.472	0.347	1.653
16	0.212	3.532	0.363	1.637
17	0.203	3.588	0.378	1.622
18	0.194	3.640	0.391	1.608
19	0.187	3.689	0.403	1.597
20	0.180	3.735	0.415	1.585

The data can be tabulated and plotted on regular paper or special chart forms can be used to make organization of the data easier. The control limits of the $\bar{x}$ chart ($\pm A_2 \bar{R}$) represent the $\pm 3\sigma_{\bar{x}}$ limits, where $\sigma_{\bar{x}}$ is the standard deviation of the averages. Since we have a relationship between $\sigma_{\bar{x}}$ and σ_x (standard deviation of the individual data), see Eq. 4.7 of Chapter 4, we can express σ_x as a function of A_2 and $\bar{R}$. This equation is:

$$\sigma_x = \frac{A_2 \sqrt{n}}{3} \bar{R} \qquad (7.10)$$

Another equation that is commonly used is:

$$\sigma_x = \frac{\bar{R}}{d_2} \qquad (7.11)$$

where d_2 is shown in Table 7.2.

From Eqs. 7.10 and 7.11 it is clear that $d_2 = 3/(A_2 \sqrt{n})$. Thus if d_2 is not known, it can be easily determined from A_2 and n.

A blank form for an $\bar{x}$ and R chart for subgroups of 5 is shown in Fig. 7.1. An example of a tabulated data of the thickness of molded panels is shown in Table 7.3; the thickness is expressed in mm × 100.

The first live rows of the table show the data of all the subgroups. The sixth row shows the averages of each subgroup, and the bottom row shows the range of each subgroup. This information will be used to construct the $\bar{x}$ and R chart for these data.

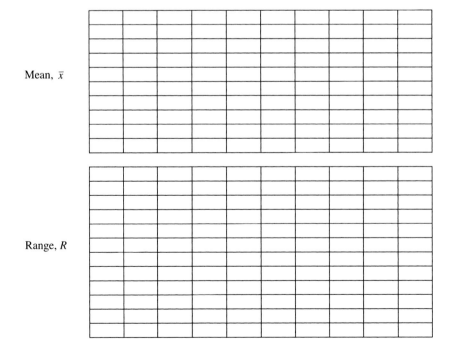

Figure 7.1 Blank form for $\bar{x}$ and R charts

Table 7.3 Tabulation of Thickness Data

	Subgroup 1	Subgroup 2	Subgroup 3	Subgroup 4	Subgroup 5	Subgroup 6	Subgroup 7	Subgroup 8
$x_1 =$	138	150	164	132	119	144	144	140
$x_2 =$	146	158	140	125	147	149	163	135
$x_3 =$	161	138	126	147	153	157	154	150
$x_4 =$	168	173	142	148	142	136	165	145
$x_5 =$	146	135	145	176	156	152	135	128
$\bar{x} =$	151.8	150.8	143.4	145.6	143.4	147.6	152.2	139.6
$R =$	30	38	38	51	37	21	30	22

For this example:

$\bar{\bar{x}} = 146.8$

$\bar{R} = 33.375$

UCL(x) = 166.06

LCL(x) = 127.54

UCL(R) = 70.55

LCL(R) = 0

The $\bar{x}$ and R charts are shown in Fig. 7.2.

Now that we know how to construct $\bar{x}$ and R charts, the next important step is the analysis and interpretation of $\bar{x}$ and R charts.

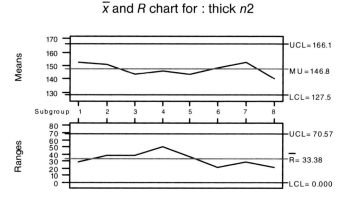

Figure 7.2 Example of $\bar{x}$ & R control chart

7.3 Interpretation of $\bar{x}$ and R Charts

The most common feature of a process showing stability is the absence of any recognizable pattern; see Fig. 7.3.

The three main characteristics of a stable process are:

- Most points are near the centerline.
- Some points are spread out and approach the control limits.
- No points are beyond the control limits.

Characteristics of an unstable process influenced by assignable causes are:

- Freaks: One or more points occur outside the control limits; see Fig. 7.4.
- Trend: There are six consecutive increases or decreases; see Fig. 7.5.
- Cycle: A repetitive pattern occurs; see Fig. 7.6.
- Run: Seven or more points in a row occur above or below the centerline; see Fig. 7.7.
- Zone Rules:
  1. Two of three successive points fall beyond two standard deviations from the mean; see Fig. 7.8, right.
  2. Four of five successive points fall beyond one standard deviation from the mean; see Fig. 7.8, left.

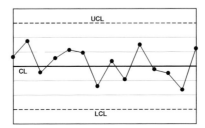

Figure 7.3 Example of a control chart of a stable process

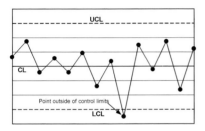

Figure 7.4 Illustration of a freak

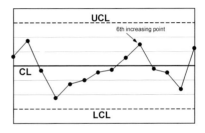

Figure 7.5 Illustration of a trend

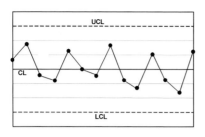 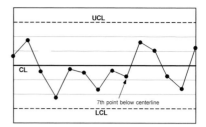

Figure 7.6 Illustration of a repetitive pattern   **Figure 7.7** Illustration of a run

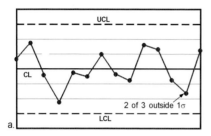 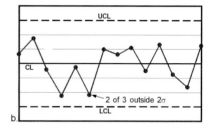

Figure 7.8 Illustrations of the zone rules, four of five outside 1σ (left), two of three outside 2σ (right)

- Shift in Level: A sudden change occurs in the central tendency of the data; see Fig. 7.9.
- Stratification: 15 or more points very close to the centerline; see Fig. 7.10.
- Clusters: Points are grouped in one area of the chart; see Fig. 7.11.
- Mixture: This is identified by an absence of points near the centerline; see Fig. 7.12.

To identify assignable causes it is best to focus first on the R chart. The R chart is more sensitive to changes in uniformity and consistency. Any change in the process will tend to shift the R values upward. Since R charts are more sensitive to changes in the process, the effect of process improvement will be noticed first in the R chart.

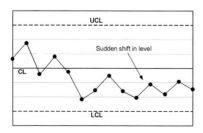 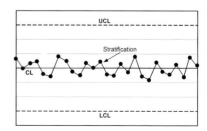

Figure 7.9 Example of a shift in level   **Figure 7.10** Example of stratification

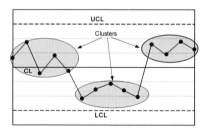

Figure 7.11 Example of clusters

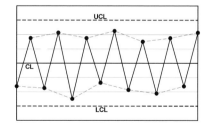

Figure 7.12 Example of mixture

The $\bar{x}$ chart should be analyzed after the R chart is stable. The $\bar{x}$ chart can be misleading when the R chart is unstable. When both the $\bar{x}$ and R charts are stable, the process is said to be in control. A stable process will have points randomly distributed between the control limits; the $\bar{x}$ and R points should tend not to follow each other. A lack of stability can cause them to move together. A process whose population is positively skewed will cause a positive correlation between the $\bar{x}$ chart and the R chart; see Fig. 7.13. Conversely, a negatively skewed distribution will cause a negative correlation between the $\bar{x}$ chart and the R chart; see Fig. 7.14. Examples of distributions with positive and negative skew are shown in Chapter 4.

SPC provides various tools to identify assignable causes of variation. However, the fact that common causes of variation are inherent to the process does not mean that they cannot be reduced or eliminated. Unfortunately, SPC does not help us much in this area. To understand the common causes of variation in a process we *have* to understand the process

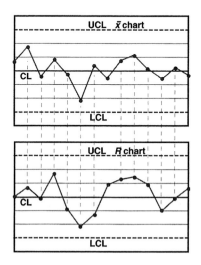

Figure 7.13 $\bar{x}$ and R with positive correlation

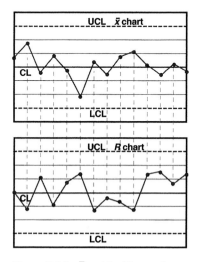

Figure 7.14 $\bar{x}$ and R with negative correlation

technology and use other statistical tools such as design of experiments (DOE). With a proper understanding of the process and with use of DOE, the effect of both assignable and common causes of variation can be analyzed and reduced to a minimum.

7.4 Control Charts for Attributes Data

Attributes data are a count of the number of nonconformities on a unit (number of defects per part) or the number of nonconforming units (defective parts). The following definitions are often used with attributes data:

Nonconformity (Defect): Any aspect or fault that causes an item (production unit or product) not to conform to the specified criteria.

c = number of occurrences of nonconformities in a sample

$u = c/n$ = number of occurrences of nonconformities per production unit

n = number of production units in the sample

Nonconforming Item (Defective): An item that does not meet measurement specifications or that contains one or more nonconformities (defects).

np = number of nonconforming items in a sample of n items

$p = np/n$ = proportion of nonconforming items in a sample

There are four types of control charts for attributes. Control charts for the count of nonconforming units are p charts and np charts. For p charts the sample size varies, for np charts the sample size is constant. Control charts for the count of individual nonconformities or defects on a product are c charts and u charts. For u charts the sample size varies, for c charts the sample size is constant. This is shown in Table 7.4.

It should be noted that the calculation of the three sigma control limits for attributes is different from that of variables. For variables it is based on the normal (Gaussian) distribution. For p charts and np charts it is based on the binomial distribution. For c charts and u charts it is based on the Poisson distribution.

Table 7.4 Control Charts for Attributes Data

Sample or subgroup size	What is counted	Type of control chart
Variable (> 50)	Defective items	p chart
Constant (> 50)	Defective items	np chart
Constant	Defects on an item	c chart
Variable	Defects on an item	u chart

7.4.1 Creating and Analyzing Attributes Control Charts

As with the $\bar{x}$ and R chart, the creation and analysis of attributes charts requires several steps. These are:

Step 1: Select the subgroup period and size. The most effective subgroup size for the *p* and *np* charts is greater than fifty. For *c* and *u* charts the subgroup size should be at least one, better between five and ten. Subgroups should be chosen to ensure minimum variation within a subgroup.

Step 2: Record the data on the control chart. Within a subgroup, items being inspected should be collected randomly so that each item has an equal chance of having nonconformities.

Step 3: Plot the data.

Step 4: Calculate and draw the control lines. At least twenty-five subgroups should be used to determine the control limits.

Step 5: Analyze the chart for signs of assignable causes of variation.

Step 6: Eliminate assignable causes and create a new chart with new control limits.

Step 7: If the process is considered in control, the control limits can be extended forward in time. As a process continues to improve, the old control limits may become too wide. In this case it can be desirable to determine new control limits based on the most recent data. The narrower the control limits can be made, the higher the quality.

7.4.2 The *p* Chart

The *p* chart (percent defective chart) is an attributes chart for the percentage of defective items in a subgroup when the subgroup size is not necessarily constant. The following symbols are used on a *p* chart:

n = number of items in a subgroup

k = number of subgroups in the study period

np = number of defective items in a subgroup

p = fraction defective in a subgroup

These are the points that are plotted on the chart. The fraction *p* is determined from:

$$p = \frac{np}{n} \qquad (7.12)$$

$\bar{p}$ is the average fraction defective for the study period. This is the central line on the chart. It is determined from:

$$\bar{p} = \frac{np_1 + np_2 + np_3 + \ldots + np_k}{n_1 + n_2 + n_3 + \ldots + n_k} \tag{7.13}$$

Since the value of n varies from subgroup to subgroup, the central limits have to be calculated for each subgroup, which makes the chart difficult to make and to read. This can be avoided by keeping the subgroup size the same or within plus or minus 25% of the average sample size and using an average n. It is also possible to show two sets of control limits based on the minimum and maximum subgroup sizes.

The upper control limit is determined using the following equation:

$$\mathrm{UCL}_{\bar{p}} = \bar{p} + 3\sqrt{\frac{\bar{p}(1-\bar{p})}{\bar{n}}} \tag{7.14}$$

where $\bar{n}$ is the average number of items in a subgroup.

The lower control limit is determined from:

$$\mathrm{LCL}_{\bar{p}} = \bar{p} - 3\sqrt{\frac{\bar{p}(1-\bar{p})}{\bar{n}}} \tag{7.15}$$

An example of a p chart is shown in Fig. 7.15.

Here, the percent defective molded pieces of connector tube is plotted for a 25 day period. The tabular data are shown in Table 7.5. For this example the average number of items in a subgroup is:

$$\bar{n} = \frac{2498}{25} = 100$$

The average fraction defective is:

$$\bar{p} = \frac{46}{2498} = 0.0184$$

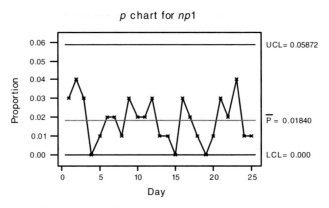

Figure 7.15 Example of a p chart for defective molded pieces

Table 7.5 Data on Defective Molded Connectors Tubes

Day	Subgroup size, n	Defectives, np	Fraction defective
1	95	3	0.0316
2	102	4	0.0392
3	108	3	0.0278
4	90	0	0.0000
5	104	1	0.0096
6	112	2	0.0179
7	88	2	0.0273
8	96	1	0.0104
9	105	3	0.0286
10	104	2	0.0192
11	92	2	0.0217
12	109	3	0.0275
13	101	1	0.0099
14	97	1	0.0103
15	93	0	0.0000
16	96	3	0.0312
17	100	2	0.0200
18	107	1	0.0093
19	103	0	0.0000
20	91	1	0.0110
21	97	3	0.0309
22	99	2	0.0202
23	108	4	0.0370
24	103	1	0.0097
25	98	1	0.0102
Sum	$\sum n = 2498$	$\sum np = 46$	

The upper control limit for the p chart is:

$$\text{UCL} = 0.184 + 3\sqrt{\frac{0.0184(1-0.0184)}{100}} = 0.0587$$

The lower control limit for the p chart is:

$$\text{LCL} = 0.184 - 3\sqrt{\frac{0.0184(1-0.0184)}{100}} = -0.0219 \quad \text{(take zero)}$$

The methods for finding signs of assignable causes are basically the same as those for $\bar{x}$ and R charts discussed earlier. The p chart in Fig. 7.15 shows no clear signs of assignable causes, which indicates that the process is in control.

7.4.3 The *np* Chart

The *np* chart is a chart for the actual number of defective items in the subgroup. It requires a constant subgroup size. The average number of defective or nonconforming units is the central line on the chart. It is determined from:

$$\overline{np} = \frac{np_1 + np_2 + np_3 + \ldots + np_k}{k} \tag{7.16}$$

where k is the number of subgroups.

The upper control limit for the *np* chart is determined from:

$$\text{UCL}_{np} = \overline{np} + 3\sqrt{\overline{np}(1-\overline{p})} \tag{7.17}$$

where: $\overline{p} - \overline{np}/n$

The lower control limit for the *np* chart is obtained from:

$$\text{LCL}_{np} = \overline{np} - 3\sqrt{\overline{np}(1-\overline{p})} \tag{7.18}$$

An example of a *np* chart is shown in Fig. 7.16.

Here, the number of defective pieces of molded caps is plotted over a 24 hour period. The tabular data are shown in Table 7.6. For this example, the average number of defective units is:

$$\overline{np} = 102/24 = 4.25$$

The average fraction of defective units is:

$$\overline{p} = 4.25/60 = 0.0708$$

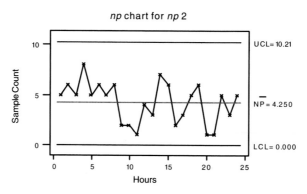

Figure 7.16 Example of an *np* chart of defective pieces of molded caps

The upper control limit is:

$$\text{UCL}_{np} = 4.25 + 3\sqrt{4.25(1-0.0708)} = 10.21$$

The lower control limit is:

$$\text{LCL}_{np} = 4.25 - 3\sqrt{4.25(1-0.0708)} = -1.71 \text{ (take zero)}$$

Figure 7.16 shows no clear signs of assignable causes.

Table 7.6 Data on Defective Pieces of Molded Caps

Hour	Subgroup size, n	Number defectives, np
1	60	5
2	60	6
3	60	5
4	60	8
5	60	5
6	60	6
7	60	5
8	60	6
9	60	2
10	60	2
11	60	1
12	60	4
13	60	3
14	60	7
15	60	6
16	60	2
17	60	3
18	60	5
19	60	6
20	60	1
21	60	1
22	60	5
23	60	3
24	60	5
		$\Sigma np = 102$

7.4.4 The c Chart

The c chart shows the number of defects or nonconformities. It is useful in situations where a unit can contain many defects. The c chart requires a constant subgroup size. It is applied where the defects are scattered continuously throughout the unit output, such as defects in a molded part or a roll of film. Subgroups can be a molded part or a square foot of film. The average number of defects for the study period is determined from:

$$\bar{c} = \frac{c_1 + c_2 + c_3 + \ldots + c_k}{k} \quad (7.19)$$

where k is the number of subgroups.

The upper control limit is obtained from:

$$\mathrm{UCL}_c = \bar{c} + 3\sqrt{\bar{c}} \quad (7.20)$$

The lower control limit is determined from:

$$\mathrm{LCL}_c = \bar{c} - 3\sqrt{\bar{c}} \quad (7.21)$$

As an example, we will consider a molding operation where one molded part is examined every thousand parts and the total number of flaws per part is recorded and graphed. The tabular data are shown in Table 7.7. For this example, the average number of defects is:

$$\bar{c} = 125/25 = 5.00$$

The upper control limit is:

Table 7.7 Flaws in Molded Parts

Number of parts	Number of flaws, c	Number of parts	Number of flaws, c
1,000	8	14,000	1
2,000	7	15,000	4
3,000	10	16,000	5
4,000	6	17,000	3
5,000	5	18,000	10
6,000	8	19,000	3
7,000	6	20,000	2
8,000	7	21,000	4
9,000	4	22,000	6
10,000	2	23,000	3
11,000	4	24,000	7
12,000	2	25,000	1
13,000	7		$\Sigma c = 125$

$$\text{UCL}_c = 5.00 + 3\sqrt{5.00} = 11.71$$

The lower control limit is:

$$\text{LCL}_c = 5.00 - 3\sqrt{5.00} = -1.71 \quad \text{(take zero)}$$

Figure 7.17 shows the c chart for the number of flaws in molded parts in Table 7.7. The c chart in Fig. 7.17 shows no clear signs of special causes.

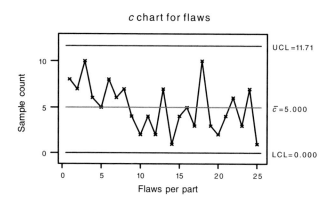

Figure 7.17 The c chart of number of flaws per molded part

7.4.5 The u Chart

The u chart can be used in the same situation as the c chart. It must be used when the subgroup size varies. The average number of defects per subgroup is:

$$u = c/n \tag{7.22}$$

where n is the number of units in a subgroup.

The grand average number of defects per unit is:

$$\bar{u} = \frac{c_1 + c_2 + c_3 + \ldots + c_k}{n_1 + n_2 + n_3 + \ldots + n_k} \tag{7.23}$$

where k is the number of subgroups.

The upper control limit is determined from:

$$\text{UCL}_{\bar{u}} = \bar{u} + 3\sqrt{\frac{\bar{u}}{\bar{n}}} \tag{7.24}$$

where $\bar{n}$ is the average number of units in a subgroup.

The lower control limit is:

$$\text{LCL}_{\bar{u}} = \bar{u} - 3\sqrt{\frac{\bar{u}}{n}} \qquad (7.25)$$

Table 7.8 shows an example of molded caps. The number of pieces, n, inspected varies from day to day. As long as n does not vary more than 25% from the average n, the average sample size can be used. The tabular data are shown in Table 7.8. For this example, the grand average number of defects per unit is:

$$\bar{u} = 80/242 = 0.331$$

Table 7.8 Actual Data of Number of Lumps in Pieces of Molded Caps

Day	Sample size, n	Number of lumps, c	Defects in subgroup, u
1	10	3	0.300
2	8	5	0.625
3	9	1	0.111
4	11	4	0.364
5	8	3	0.375
6	9	3	0.333
7	12	1	0.083
8	8	6	0.750
9	9	5	0.556
10	9	4	0.444
11	11	3	0.273
12	10	3	0.300
13	11	1	0.091
14	9	2	0.222
15	10	5	0.500
16	8	4	0.500
17	9	4	0.444
18	10	7	0.700
19	9	3	0.333
20	11	1	0.091
21	11	3	0.273
22	9	4	0.444
23	10	1	0.100
24	12	2	0.167
25	9	2	0.222
	$\Sigma n = 242$	$\Sigma n = 80$	

and the average sample size is:

$$\bar{n} = 242/25 = 9.68$$

We will take the closest whole number for the average sample size; thus the sample size is $\bar{n} = 10$. The upper control limit thus becomes:

$$UCL_u = 0.331 + 3\sqrt{\frac{0.331}{10.0}} = 0.877$$

The lower control limit is:

$$UCL_u = 0.331 - 3\sqrt{\frac{0.331}{10.0}} = -0.215 \quad \text{(take zero)}$$

Figure 7.18 shows a u chart for the number of lumps in pieces of molded caps.

The control chart shows no clear signs of special causes of variation. The upper control limit in the chart is 0.8567, while UCL = 0.877 according to the calculation using Eq. 7.24. The reason for this discrepancy is that we rounded the average sample size to $\bar{n} = 10$. As a result, the UCL is a little lower in the chart as is the central line, $\bar{u}$. The central line in the chart is actually calculated from $\bar{u} = \sum c / k\bar{n}$.

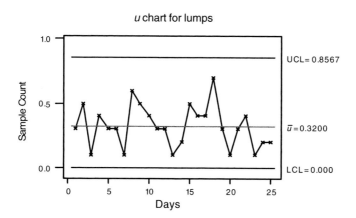

Figure 7.18 The u chart for lumps in molded caps

This problem can be avoided by not using an average sample size, but the actual sample sizes. This, of course, means that the upper control limit will vary as the sample size varies. Thus the upper control limit will not be a straight line, but a stepped line instead. The u chart using the actual sample sizes is shown in Fig. 7.19. Notice that the central line now has the correct value, $\bar{u} = 0.331$ (actually 0.3305785124). The value of the upper control limit shown on the u chart is the highest value.

8   Process Capability and Special SPC Techniques for Molding and Extrusion

8.1   Introduction

Process capability is a measure of how capable a process is of making parts that are within specifications. Figure 8.1 shows a capable process where the spread of the process data is narrower than the tolerance or specification width.

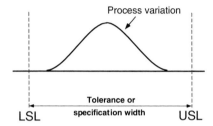

Figure 8.1   Illustration of a capable process

In other words, all the process data shown in the figure are within the upper and lower specification limits. Before the process capability is determined, one has to make sure that the process is in control. This is done by making sure that no assignable causes are affecting the process. Thus if assignable causes are acting on the process, these have to be eliminated first. In determining process capability we need to determine the spread of the individual process data (x), not of the averages $(\bar{x})$. Therefore, we cannot simply use the control limits from an $\bar{x}$ chart, because the control limits describe the $6\sigma_{\bar{x}}$ spread of $\bar{x}$. As discussed earlier, the standard deviation of the individual data, σ_x, is related to the standard deviation of the averages $\sigma_{\bar{x}}$ by:

$$\sigma_x = \sigma_{\bar{x}} \sqrt{n} \tag{8.1}$$

where n is the number of data from which $\bar{x}$ is calculated (the size of the subgroup). The spread of the individual data is always larger than the spread of the averages; see Fig. 8.2.

Process capability is determined to find out whether a process can meet specifications and how many parts can be expected to be out of tolerance. Capability studies can also be useful in the following situations:

- Evaluation of new equipment purchases
- Equipment selection for production of certain parts
- Setting specifications
- Costing out contracts

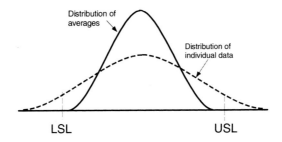

Figure 8.2 Distribution of the averages and the individual data

The steps involved in a capability study are the following:

1. Collect process data and make sure measurement system is capable.
2. Plot the data on control charts.
3. Determine the control limits.
4. Make sure the process is in control.
5. Determine the process capability.
6. If the process is not capable, improve the process and return to step 1.

Before process capability is determined, one has to make sure that the process is *in control* and that the individual process data is *normally distributed*. It should be noted that a process can be in control and yet have its individual process data not normally distributed. This happens when the process is inherently nonnormal. Certain measurements will always result in nonnormal distributions. Examples are roundness, flatness, and so on; these measurements have a natural barrier at zero. A perfect measurement is zero; values less than zero are not possible. Standard capability indices are not valid for such nonnormal distributions.

8.2   Capability Indices

The most commonly used capability indices are the process potential index, CP, and the capability index CpK. The process potential index is the ratio of the tolerance to six times the process standard deviation of the *individual* data, σ_x:

$$\text{CP} = \frac{\text{Tolerance}}{6\sigma_x} \qquad (8.2)$$

where: tolerance = USL−LSL.

The acronym USL stands for the upper specification limit and LSL is the lower specification limit. The standard deviation of the individual data, σ_x, is often determined from $\bar{x}$ control limits of the $\bar{x}$ and R chart; this involves dividing the average range by a constant d_2, see Chapter 7, Eq. 7.11. In some cases, the σ_x is determined from the sample standard deviation, for instance, when we use an $\bar{x}$ and s chart. When the σ_x is determined from the sample standard deviation, the ratio of tolerance to $6\sigma_x$ is also referred to as the performance index or Pp. Unfortunately, there is a lack of uniformity in the nomenclature used for process capability indices. As a result, one should always verify exactly what definition of the capability index is used.

In a normal distribution, 99.73% of the data will be within a $6\sigma_x$ spread, see Chapter 4, Fig. 4.19. Thus when CP = 1.0, it can be expected that 99.73% of the parts will be within the tolerance (assuming that the data are centered around the target value). The higher the value of CP, the more parts will fall within the specification limits. For example, when CP = 1.33, it can be expected that 99.994% of the parts will be within tolerance. In other words, only 6 out of 100,000 parts would be expected to be out of tolerance. When CP = 1.0, we would expect 270 out of 100,000 parts to be out of tolerance. Thus a small change in CP can translate in a large change in the number of rejects.

CP is only a measure of spread of the process data. It does not take into account whether or not the data are centered around the target value. If the data are not centered around the target value, using the CP can result in a false sense of security. In this case, the CpK is more useful because it takes both the spread and the center of the data into account. The equation for CpK is:

$$\text{CpK} = \frac{\text{USL} - \text{MEAN}}{3\sigma_x} \quad (8.3a)$$

or

$$\text{CpK} = \frac{\text{USL} - \text{LSL}}{3\sigma_x} \quad (8.3b)$$

The CpK is the lowest of these two values. If the mean is larger than the target value, Eq. 8.3a will yield the lowest CpK. If the mean is smaller than the target value, Eq. 8.3b will give the lowest CpK. The value of the CpK from Eq. 8.3a is sometimes referred to as the upper capability index or CPU; conversely, the value from Eq. 8.3b is referred to as lower capability index or CPL.

If the data are centered around the target value, the CP = CpK. However, when the data are not centered around the target value, the CpK will always be smaller than the CP. Figure 8.3 shows examples of CP and CpK values for four different distributions.

Many companies require a CpK index of 1.33 to 2.0 for the product from their suppliers. A CpK of 1.33 means that 6 or fewer parts per 100,000 are expected to be out of specifications. A CpK of 2.0 means that 2 or fewer parts per 1,000,000,000 are expected to be out of specification. Clearly, a CpK of 2.0 represents an almost negligible chance of rejects.

8 Process Capability and Special SPC Techniques

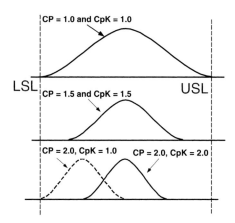

Figure 8.3 CP snd CpK for different distributions

Sometimes another capability index is used; this is the CR index or capability ratio. The CR index is simply the inverse of CP:

$$CR = \frac{6\sigma_x}{\text{tolerance}} \qquad (8.4)$$

The smaller the CR value, the better. The CP index is used more often than the CR index, because it is easier to compare to CpK. Sometimes not only the lowest CpK value is used from Eqs. 8.3a and 8.3b, but also the largest. The larger of the two CpK values is also referred to as the Zmax/3 index. Using the same nomenclature, the CpK index can also be called the Zmin/3 index.

8.2.1 Process Capability Shortcut with Precontrol

It is clear that determining process capability using conventional SPC is a rather time-consuming process. In the precontrol method of SPC, the process capability is determined simply by taking five consecutive units from the process. Precontrol will be discussed in Chapter 9. In precontrol, the tolerance is divided into two zones: the middle zone is called the green zone and the zones between the green zone and the specification limits are the yellow zones. The regions outside the specification limits are the red zones, see Fig. 8.4.

If five consecutive units fall within the green zone, the process is considered capable. It can be shown that the capability index is at least 1.33. In conventional SPC, it takes at least fifty units from the process before the process capability can be determined.

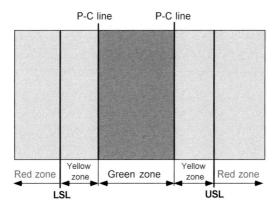

Figure 8.4 The three color zones in precontrol

8.3 Tests for Normality

The capability indices CP, CpK, and CR all assume normal distribution of the data. Fortunately, the above equations can be used for a number of nonnormal distributions without introducing much error [25]. Even in the highly nonsymmetrical exponential distribution, the $\pm 3\sigma_x$ limits will contain 98.2% of the data, as opposed to 99.7% for the normal distribution. Strictly speaking, we have to determine the distribution of the data when we determine process capability. A number of tests can be used to determine whether or not the distribution is normal. These tests can be divided into two categories: graphical methods and statistical calculations.

8.3.1 Graphical Methods

If the sample is large enough, a histogram can be plotted. Then a normal curve with the same mean and standard deviation can be plotted with the histogram to see how well the normal curve fits the actual distribution. This is not a quantitative comparison and, therefore, not a good test of normality.

Another graphical method is a plot of the cumulated distribution an normal probability paper to determine how well it is fitted by a straight line. A straight line indicates a normal distribution. This plot is usually referred to as a normal probability plot. Figure 8.5 shows an example of a normal probability plot of the thickness data covered in Chapter 7, Table 7.3.

There is a good fit to a straight line, so it would be safe to assume that the distribution of the data is close to normal. The straightness of the probability plot can be measured by the correlation coefficient. An efficient test for normality, which is essentially equivalent to the

Shapiro-Wilk test [39] can be based an this correlation. A very high correlation is consistent with normality. Critical values for the correlation are shown in Table 8.1.

The hypothesis of normality is rejected if the correlation falls below the critical values indicated in Table 8.1.

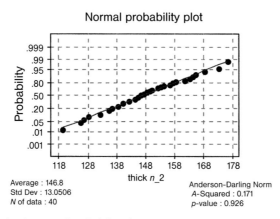

Figure 8.5 Example of a normal probability plot

Table 8.1 Correlation Test for Normality

	alpha		
N	0.10	0.05	0.01
4	0.8951	0.8764	0.8318
5	0.9033	0.8804	0.8320
10	0.9349	0.9176	0.8792
15	0.9506	0.9383	0.9110
20	0.9602	0.9511	0.9270
25	0.9662	0.9582	0.9408
30	0.9707	0.9639	0.9490
40	0.9769	0.9717	0.9579
50	0.9807	0.9764	0.9664
60	0.9835	0.9799	0.9710
75	0.9865	0.9835	0.9757
80	0.9871	0.9843	0.9776
100	0.9894	0.9871	0.9818
400	0.9969	0.9964	0.9950
600	0.9979	0.9975	0.9966
1000	0.9987	0.9984	0.9979

8.3.2 Statistical Calculations

A commonly used test of normality is the chi-square, χ^2, test. This test is also called goodness-of-fit test, because we determine how well the normal curve fits our data. The data are divided into classes (intervals). Each class has an upper and lower boundary and we determine how many data points fall in each class. If there are not at least five data points in a class, it is combined with the next class. Then the chi-square calculation is performed for each class.

The chi-square formula is:

$$\chi^2 = \sum \frac{(O-E)^2}{E} \tag{8.5}$$

Where O is the observed frequency and E the expected frequency based on the normal distribution. The Σ means that the $(O-E)^2/E$ values for all classes are summed to obtain chi-square. This value is compared to the value of the chi-square in Table 8.2.

Table 8.2 The Chi-Square Distribution Table

DF	χ^2 .005	χ^2 .01	χ^2 .05	χ^2 .10	χ^2 .90	χ^2 .95	χ^2 .99	χ^2 .995
1	39E-6	16E-5	39E-4	0.016	2.71	3.84	6.63	7.88
2	0.010	0.020	0.103	0.210	4.61	5.99	9.21	10.60
3	0.072	0.115	0.352	0.584	6.25	7.81	11.34	12.84
4	0.207	0.297	0.711	1.064	7.78	9.49	13.28	14.86
5	0.412	0.554	1.15	1.61	9.24	11.07	15.09	16.75
6	0.676	0.872	1.64	2.20	10.64	12.59	16.81	18.55
7	0.989	1.24	2.17	2.83	12.02	14.07	18.48	20.28
8	1.34	1.65	2.73	3.49	13.36	15.51	20.09	21.96
9	1.73	2.09	3.33	4.17	14.68	16.92	21.67	23.59
10	2.16	2.56	3.94	4.87	15.99	18.31	23.21	25.19
11	2.60	3.05	4.57	5.58	17.28	19.68	24.73	26.76
12	3.07	3.57	5.23	6.30	18.55	21.03	26.22	28.30
13	3.57	4.11	5.89	7.04	19.81	22.36	27.69	29.82
14	4.07	4.66	6.57	7.79	21.06	23.68	29.14	31.32
15	4.60	5.23	7.26	8.55	22.31	25.00	30.58	32.80
16	5.14	5.81	7.96	9.31	23.54	26.30	32.00	34.27
18	6.26	7.01	9.39	10.86	25.99	28.87	34.81	37.16
20	7.43	8.26	10.85	12.44	28.41	31.41	37.57	40.00
24	9.89	10.86	13.85	15.66	33.20	36.42	42.98	45.56
30	13.79	14.95	18.49	20.60	40.26	43.77	50.89	53.67
40	20.71	22.16	26.51	29.05	51.81	55.76	63.69	66.77
60	35.53	37.48	43.19	46.46	74.40	79.08	88.38	91.95
120	83.85	86.92	95.70	100.6	140.2	146.6	158.9	163.6

DF stands for the number of degrees of freedom. It is determined by the number of classes minus three. We subtract three because we are estimating three statistics: mean, standard deviation, and frequency. The different columns represent different risk factors. Typically, a risk factor of 0.95 is used. The value from the table is called the critical chi-square. If the calculated chi-square value is smaller than the critical value based on a risk factor of 0.95, we can say that the distribution is normal with a better than 95% confidence level. Further information on the chi-square test can be found in [3].

Other calculations to test for normality determine certain measures of nonnormal characteristics such as skewness and kurtosis. If a distribution is symmetrical (zero skewness), its third moment about the mean will be zero. The third moment or skewness is a measure of how much heavier the distribution is on one side versus the other. It is determined from:

$$M_3 = \frac{\sum (x_i - \bar{x})^3}{n-1} \tag{8.6}$$

If the distribution has a long tail to the right, the third moment about the mean will have a positive value. A negatively skewed distribution will have a negative value of the third moment about the mean. Figure 8.6 shows positively and negatively skewed distributions.

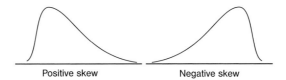

Figure 8.6 Examples of distributions with positive and negative skew

Skewness may be estimated by the difference between the mean and the mode divided by the standard deviation:

$$M_3 \approx \frac{\bar{x} - \hat{x}}{s} \tag{8.7}$$

Further information can be found in Duncan's book [3].

A measure of kurtosis is the ratio of the fourth moment to the second moment about the mean; the fourth moment about the mean is determined from:

$$M_4 = \frac{\sum (x_i - \bar{x})^4}{n-1} \tag{8.8}$$

For a normal distribution this M_4/M_2 ratio has a value of 3. When the ratio is larger than 3, the distribution has a sharper peak and fatter tails than the normal distribution. When the ratio is less than 3, the distribution has a flatter peak and thinner tails than the normal distribution.

8.4 Capability Measures for Non-Normal Distributions

If the distribution that we are dealing with is not normal, there are various approaches to determine the capability of the process. In some cases the nonnormal distribution can be transformed to a normal distribution. This is usually done by taking the natural logarithm of the data, including the specification limits. If the distribution has a natural barrier at zero, the process capability can be expressed as the percentage of parts between zero and the upper specification limit. Another option is to calculate the CpK for just the upper half of the distribution.

If the distribution is unimodal but not symmetrical, we can split the distribution data in two halves. The central point should be determined by the mode. Each half is plotted separately and its mirror image added to give two approximately normal distributions. Each distribution curve can then be analyzed for the percentage of parts outside the appropriate specification limit. We should calculate the area below LSL using curve I and the area above USL using curve II; see Fig. 8.7.

Since we are only concerned with half of each distribution, the area within the curve is determined by subtracting the area outside the specification limit from 0.50. The total percentage of the actual distribution within specification limits is found by adding the results of half of curve I and half of curve II. This total percentage can be used to quantify the process capability. A condition for this approach, obviously, is that both halves of the curves can be fitted reasonably well with a normal curve.

Several other methods to determine process capability for nonnormal distributions are available. One is based one the Weibull distribution. Another useful method is the use of statistical tolerances according to Wilks [10]. This approach is valid for any distribution as long as the variable is continuous and the sampling random. Wilks derived an expression from which one can calculate the sample size n necessary for an assurance A that at least $100B\%$ of the contents of a lot or the output of a random process will be included between the largest and smallest values in a random sample from the lot or process:

$$nB^{n-1} - (n-1)B^n = 1 - A \tag{8.9}$$

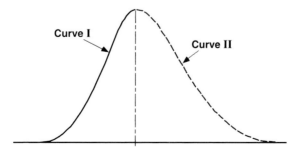

Figure 8.7 Splitting of a unimodal but asymmetrical distribution

The problem with the Wilks method is that for a high level of assurance, for example, for $A = 0.95$, and a high value of B, for example $100B = 99.73\%$, the sample size becomes quite large. For $A = 0.95$ and $B = 0.9973$, the necessary sample size is 1755. This method, therefore, may not be practical if the measurements are made by hand. However, if measurements are made automatically and on-line, such as a diameter measurement using a laser gage, then a large sample size is no major problem. Even if the data are collected only once per second, then a sample of 1755 measurements will be collected within half an hour.

Clements [15] proposed a simple method for calculating CP and CpK for any shape distribution, using the Pearson family of curves. The calculation requires estimates of the mean, standard deviation, and kurtosis for the process. The advantages of this method are:

- When the distribution is normal the indexes are exactly the same as those given by the traditional method.
- The only difference from the traditional procedure is the method of calculating the width and position of the top and bottom halves of the distribution relative to the tolerance limits.
- It does not require mathematical transformation of the data.
- It is easy to visualize graphically and explain to nonstatisticians.
- It is relatively easy to calculate manually or by using a handheld calculator.

One of the most convenient ways of doing SPC is by using personal computers. There is a great multitude of good software commercially available that makes doing SPC much easier

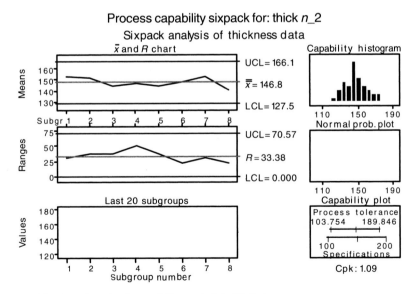

Figure 8.8 The sixpack option in capability analysis in Minitab

and quicker. Some of the SPC software packages are listed in Appendix III. An example of a quick overview of a substantial amount of SPC information is shown in Fig. 8.8.

The sixpack feature shows the $\bar{x}$ and R chart, the actual data of the last 20 subgroups, the capability histogram of the data, the normal probability plot (as shown also in Fig. 8.5), and a capability plot showing how the process variation compares to the tolerance. Once the data are entered into a spreadsheet format, generating the plots shown in Fig. 8.8 takes less than 15 seconds. More detailed information on capability analysis is shown in Fig. 8.9.

Figure 8.9 shows the Cp, CPU, CPL, CpK, USL, LSL, k, n, mean, mean + 3s, mean − 3s, and so on. Figure 8.10 was also generated using Minitab. When having to determine process capability with nonnormal data, a good option is to use a SPC software package that can handle nonnormal data. In fact, this would probably be the most expedient and accurate way to handle the nonnormal data analysis.

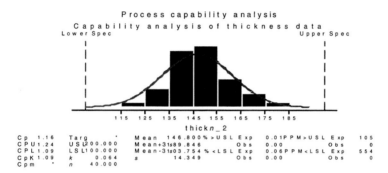

Figure 8.9 Process capability information on the thickness data

8.5 Special SPC Techniques for Injection Molding

Since injection molding poses some special problems in SPC, as opposed to a continuous process like extrusion, some special techniques have been developed especially for the injection molding process. These will be discussed next.

8.5.1 Family Processes

A family process consists of several statistically independent processes that are affected by common factors; these processes are also referred to as "multi-stream" processes. An example of a family process is injection molding using a multicavity mold. The filling and cooling in some cavities may be affected by factors not acting on the other cavities. If an operator samples live parts from a thirty-two cavity mold, the probability of a cavity not being included in the sample is 83.3%. If the samples are taken on an hourly basis, production may continue for a full shift or even a day without sampling one of the cavities. Thus nonconforming parts may go undetected for a substantial period of time.

When a faulty cavity is part of a sample that leads to an out-of-control point, a common tendency is to adjust factors that affect all the cavities — these are the "global" factors. If the faulty cavity is not included in the next sample, the process may seem to have been properly adjusted. In actuality, the operator erroneously changed a process that was in statistical control.

8.5.1.1 Median/Individual Measurement Control Charts

One solution to this problem was proposed by Grant and Leavenworth in their book *Statistical Process Control* [26]. They proposed combining the results of median charts with individual measurement charts. The latter track local variations, while the former monitor global variation. One advantage of median charts is that they require no calculations by the operator.

In a median chart, each sample or observation consists of a unit from each family member. Remember that the median, $\tilde{x}$, is the point that divides the values of the individual measurements in half. The frequency of sampling is dependent on the process; initially, the sampling should be frequent enough to profile the process. If the process is stable, then the sampling frequency can be reduced. Determination of the average median is similar to that of the grand average of a $\bar{x}$ chart:

$$\bar{\tilde{x}} = \frac{\tilde{x}_1 + \tilde{x}_2 + \tilde{x}_3 + \ldots + \tilde{x}_k}{k} \tag{8.10}$$

where k is the number of subgroups.

The upper control limit is determined by:

$$\text{UCL}_{\tilde{x}} = \bar{\tilde{x}} + \tilde{A}_2 \bar{R} \tag{8.11}$$

The lower control limit is determined by:

$$\text{LCL}_{\tilde{x}} = \bar{\tilde{x}} - \tilde{A}_2 \bar{R} \tag{8.12}$$

Factor $\tilde{A}_2$ is given in Table 8.3; note that the value of $\tilde{A}_2$ is different from the value A_2 for $\bar{x}$ and R charts given in Chapter 7, Table 7.2. The control limits for the individual measurement, x, are determined from:

$$\text{UIL}_x = \bar{\tilde{x}} + E_2 \bar{R} \tag{8.13}$$

The lower individual control limit (LIL) is determined from:

$$\text{LIL}_x = \bar{\tilde{x}} - E_2 \bar{R} \tag{8.14}$$

The factors $\tilde{A}_2$ and E_2 are given in Table 8.3.

8.5 Process Capability and Special SPC Techniques

Table 8.3 Factors with the Median ($\tilde{A}_2$) and Individual Measurement (E_2) Chart

n	$\tilde{A}_2$	E_2
3	1.187	1.772
4	0.796	1.457
5	0.691	1.290
6	0.548	1.184
7	0.508	1.109
8	0.433	1.054
9	0.412	1.010
10	0.362	0.975

The median/individual chart shows the individuals as points and the median as a moving line; see Fig. 8.10. The measurements from one sample are shown along one vertical line; in this figure the measurements are shown as plus marks (+).

The median/individual charts for family processes are easy to use and interpret. The analysis is more efficient since the effect of global and local causes of variation can be differentiated. The overall process variation can be reduced by centering the individuals.

If the median is within its control limits and one or more individuals exceed those limits, the problem lies with the offending individuals. An example of such a situation is shown in Fig. 8.11.

In this case, global factors should not be adjusted, because it is local factors that cause the problem. Local factors could be factors such as the temperature in a particular region of the mold, the size of the runner leading to one or more cavities, the size of the gate, and so on.

If the median exceeds its limits, a change in global factors would be appropriate. An example of the median going out of control is shown in Fig. 8.12.

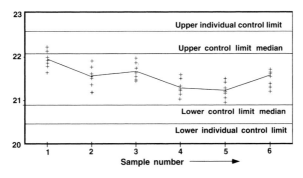

Figure 8.10 An example of a median/individual chart

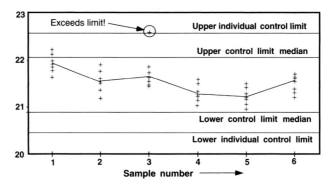

Figure 8.11 An example of a median/individual chart with an out-of-limit individual

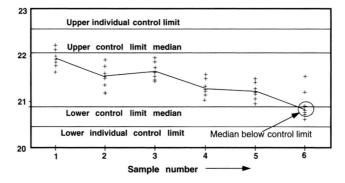

Figure 8.12 An example of a median/individual chart with median out of control limits

Van der Veen and Holst [27] describe an application of the median/individual charting technique to the molding of a sprayer body in an eight cavity water-cooled injection mold. The median chart indicated global shifts in the process. These were associated with changes in the color of the primary material. The individuals chart showed that cavity number six was consistently lower than all others. It was found that the cooling water temperature for cavity six was lower than that for the other cavities, leading to faster cooling and less efficient packing. By adjusting the water flow, this problem could be eliminated. This charting technique is included in some SPC software packages. Using a computer to generate control charts is certainly to be preferred over manual control charting; it is faster and reduces human errors, allowing people to be involved in more productive activities.

8.5.1.2 Group Charting

The problem of conventional control charting of multistream (family) processes was discussed by Bajaria and Skog [28]. One method of dealing with multicavity injection molding processes is to use x and Rm charts (individual measurement and moving range) for each individual stream; this method is recommended for cases where the CpK of the cavities is equal to or greater than 3 (CpK ≥ 3). When CpK < 3, the $\bar{x}$ and R chart (average and range) is recommended. This approach, though, tends to be complicated and time consuming.

An alternative and more expedient approach is to use the group chart. In this chart the highest and lowest values are plotted on the x chart and the largest moving range on the R chart. An example of a group chart is shown in Fig. 8.13.

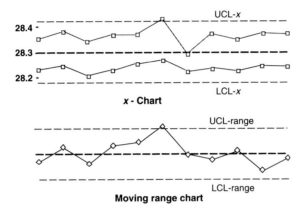

Figure 8.13 Example of a group chart

The process is running well as long as the high x value is below the upper action limit and the low x value is above the lower action limit. The moving range chart shows the maximum range from any stream to its own previous value, making it sensitive to changes in any stream regardless of constant differences between streams. The group chart is most useful for day-to-day monitoring. However, a tabular report can be useful to identify consistently high or low stream. For instance, from a tabular report it may become obvious that cavity number three is running consistently high. If we use a median/individual chart, we can denote the individual points with the cavity number. In this case, it will be immediately obvious from the chart when one of the cavities is running consistently low or high. Identifying the low and high x values with the cavity number will do the same in the group chart. This method is preferred, because it is easier to identify problems from a chart than from tabular data.

8.5.2 Grading Machine Capability

SPC can be used to assess the capability of an injection molding machine. A major study on more than 1800 injection molding machines was undertaken by Hunkar [30]. The data was compiled using PMAs (portable machine analyzer) once the machines ran on a stable cycle.

Data were gathered on nineteen parameters over the course of at least thirty-two to fifty shots, preferably one hundred. For each parameter, the mean was determined over the total number of shots and also the difference from the mean for each shot. The latter determines the classification of the machine for the various parameters studied.

The results were worked out so that injection molding machines were divided into nine classes, I being the best and IX being the worst. The information is shown in Table 8.4. Obviously, other methods are possible to assess the capability of injection molding machines. However, the data presented by Hunkar appear to be the first comprehensive study in this area and provide a clear method of assessing injection molding machine capability.

Table 8.4 shows that a class I machine can keep the hold time to a variation of less than 0.02 seconds. On the other hand, a class V molding machine can keep the hold time to a variation of less than 0.04 seconds; this is twice as much variation. A class I machine can

Table 8.4 Nine Classes of Injection Molding Machine Variations, According to Hunkar [30]

Parameter	Class								
	I	II	III	IV	V	VI	VII	VIII	IX
Cycle time, s	0.20	0.24	0.29	0.35	0.41	0.50	0.60	0.72	0.86
Hold time, s	0.02	0.02	0.03	0.03	0.04	0.05	0.06	0.07	0.09
Inject time, s	0.04	0.05	0.06	0.07	0.08	0.10	0.12	0.14	0.17
Clamp closed, s	0.10	0.12	0.14	0.17	0.21	0.25	0.30	0.36	0.43
Clamp open, s	0.10	0.12	0.14	0.17	0.21	0.25	0.30	0.36	0.43
Plasticate, s	0.15	0.18	0.22	0.26	0.31	0.37	0.45	0.54	0.64
Cavity pressure, psi	15.00	18.00	21.60	25.92	31.10	37.32	44.79	53.75	64.50
Peak injection pressure, psi	20.00	24.00	28.80	34.56	41.47	49.77	59.72	71.66	86.00
Hold pressure, psi	4.00	4.80	5.76	6.91	8.29	9.95	11.94	14.33	17.20
Backpressure, psi	5.00	6.00	7.20	8.64	10.37	12.44	14.93	17.92	21.50
Ram stroke, in	0.05	0.06	0.07	0.09	0.10	0.12	0.15	0.18	0.21
Mold A temperature, °F	3.00	3.60	4.32	5.18	6.22	7.46	8.96	10.75	12.90
Mold B temperature, °F	3.00	3.60	4.32	5.18	6.22	7.46	8.96	10.75	12.90
Oil temperature, °F	3.00	3.60	4.32	5.18	6.22	7.46	8.96	10.75	12.90
Dewpoint, volts	0.01	0.01	0.01	0.02	0.02	0.02	0.03	0.04	0.04
Temperature 1, °F	2.00	2.40	2.88	3.46	4.15	4.98	5.97	7.17	8.60
Temperature 2, °F	2.00	2.40	2.88	3.46	4.15	4.98	5.97	7.17	8.60
Temperature 3, °F	2.00	2.40	2.88	3.46	4.15	4.98	5.97	7.17	8.60
Temperature 4, °F	2.00	2.40	2.88	3.46	4.15	4.98	5.97	7.17	8.60

keep the peak injection pressure variation to less than 20.00 psi, while a class V machine can keep the peak injection pressure variation to less than 41.47 psi.

Some injection molding machine manufacturers have adopted this classification of machine capability and guarantee performance according to a certain class. Obviously, processors can use this information to determine how capable their machines are; this will provide good information to determine whether certain machines should be replaced or reconditioned. It should be noted that this machine capability classification is based on machines evaluated before 1992. As injection molding machines continue to be improved, the values that go with the different classes will change over time. As a result, Table 8.4 will have to be adjusted periodically, perhaps every five years, to reflect improvements in injection molding machines. Obviously, the changes will affect mostly the numbers in the class I machine column.

8.1 Example of SPC in Extrusion

This example describes the implementation of SPC at a profile extruder using very simple tools. A paper on this subject was presented at the 1999 SPE ANTEC in New York [49].

8.6.1 Introduction

Many custom profile extrusion companies specializing in low technology commodity extrusions are able to manufacture at very attractive prices. This price advantage, in part, comes from the low overhead required to operate these plants and the minimal demand customers have on the parts they buy. Color, consistency, tolerances, and finish are often of little consequence to the end customer. Commodity profile extrusion companies are likely to be high volume resin users and therefore also enjoy volume cost advantages.

By allowing customers knowledgeable in SPC fundamentals to bring simple, inexpensive SPC into these plants for the lines running their profiles, it is possible to increase the plants' ability to make more demanding extrusions without increasing its overhead. Lower production costs often follow better control so that both the profile extruder and customer can gain cost benefits. In the example considered here, the customer also provided some of the downstream equipment and the laser measuring sensor and display to the profile extrusion company.

For this approach to work, the potential volume of business the customer represents must be attractive to the profile extrusion company. Furthermore, the profile extrusion company must be willing to have a close working partnership with the customer rather than maintain the typical arm's-length customer-vendor relationship. Diplomacy by the customer and open-mindedness by the vendor are mandatory.

8.6.2 Specific Example Overview

A customer with a requirement for acrylic profiles had gotten quotes from several conventional custom profile extrusion companies that were well above the costs he required.

A respected commodity extruder specializing in diffusers for fluorescent light fixtures and signage extrusions was contacted, an extruder who used large quantities of acrylic. It was determined that the potential volume the customer represented could amount to roughly 4% of this profile extrusion company's current sales. This profile extrusion company was skeptical that the customer's desired costs could be met, but was willing to allow the customer to rent an extrusion line on a time and materials basis with the profile extrusion company supplying the operator.

It was agreed that, for the initial run, a single profile would be selected. The profile extrusion company would design and build the die, then set up and run the line using its normal operating practices. The customer was to make whatever measurements he felt were necessary and collect and record data where appropriate. If necessary, at the customer's expense, the line would be modified based on suggestions by the customer.

The customer brought a laser curtain (Keyence Model LS-5501, Keyence Corp., Woodcliff Lake, NJ) into the plant to measure the critical dimension of the profile on-line. This model laser curtain had substantially more capability than necessary, but was already available to the customer. A Keyence model LX-132 laser system with RS232 output could also have been used and would have been much less expensive.

The initial test run produced parts below the run speed projected by the customer, but was about what the profile extruder expected. The line produced parts with a process potential index (CP) of less than one for the profile's key dimension. Parts showed errors from possible causes such as melt fracture, draw resonance and take-up speed variation.

After applying the SPC methods discussed in this book and making appropriate changes to the line, the part extrusion speed was increased about 50%. The customer's goal of a CP of 1 or more was achieved, and no signs of problems from melt fracture, draw resonance, or take-up speed variations were apparent.

After this demonstration, the profile extrusion company agreed that it could sell parts at the cost per foot initially proposed by the customer. Parts are now being produced in production volumes. The customer's purchases have exceeded initial projections.

The customer has left the laser curtain and its display with the profile extrusion company. These data are not collected or used to control the process directly. The laser curtain's display provides a real-time graph of the key dimension as the part passes through the laser curtain. The graph shows the LSL (lower specification limit) and USL (upper specification limit) for the dimension. The sensor is programmed to set off an alarm if measurements occur outside of preset limits, that are well within the LSL and USL. If the alarm is not acknowledged within a preset time, the line is automatically shut down.

The customer makes occasional on-site visits to the plant to download data into his laptop computer and analyze it in the spreadsheets he designed for that purpose.

8.6.3 Low Cost Portable SPC Package

There are dozens of process parameters that can be measured and controlled in the extrusion process. Measuring extrudate dimensions is one of the most important and useful. It is also one of the easiest to measure on a continuous basis.

The hardware used for this project was a popular brand laptop computer running a Windows™ based spreadsheet program and an off-the-shelf standard laser based non-contact dimension measuring sensor with RS232 output (Keyence Model LS-5501). The data acquisition software was WinWedge (version 3.0) from TAL Technologies, Philadelphia, PA. This is an inexpensive (under $500) software package capable of inserting output from the measuring sensor directly into a spreadsheet.

The I/O (input/output) software capabilities far exceeded those required for this project and included the ability to control, translate, and interrogate multiple input sources and output them into almost any Windows™ based program with little or no programming experience required. The total cost for this system, not including the computer, laser gage, and spreadsheet software, was around $2000.

The major cost component for the system was the laser curtain. However, it should be noted that a much less expensive measurement system could have been used for this application. The advantage of this system was that components could be taken out of the box, plugged in, and used without the need to write any specialized software or to design and build any special hardware interfaces. This setup can be run on almost any computer capable of running Windows™ as speed and memory requirements for the I/O software are minimal.

Setting up the software to interrogate and read data from the sensor and automatically input it into designated cells in the spreadsheet took less than a day. Once this was accomplished, a spreadsheet was designed using standard functions found in all popular spreadsheets to process data from the measurement sensor into a relevant suite of statistical calculations.

Since data could be easily and accurately collected directly into the spreadsheet, it was decided that twenty subgroups of data with twenty data points each would be collected for analysis. Figure 8.14 shows an example of the spreadsheet with actual data. The following calculations were generated for each full set of data collected: average, range, standard deviation, rolling average standard deviation, rolling average, high and low z values, and estimated percentage of measurements out of tolerance for the sample. This last calculation was later converted into the estimated number of parts in the sample not meeting specifications.

All of these calculations can be made without any special knowledge of statistics or SPC, and all of the functions necessary to calculate these results are basic and standard on all

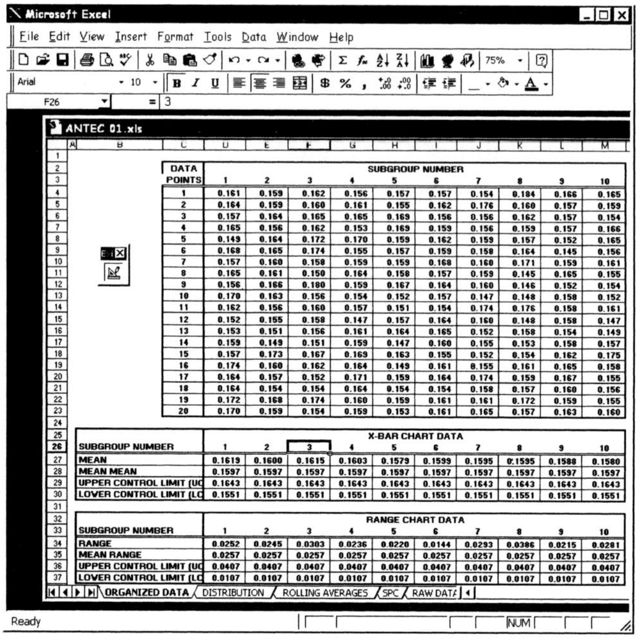

DATA POINTS	SUBGROUP NUMBER 1	2	3	4	5	6	7	8	9	10
1	0.161	0.159	0.162	0.156	0.157	0.157	0.154	0.184	0.166	0.165
2	0.164	0.159	0.160	0.161	0.155	0.162	0.176	0.160	0.157	0.159
3	0.157	0.164	0.165	0.165	0.169	0.156	0.156	0.162	0.157	0.154
4	0.165	0.156	0.162	0.153	0.169	0.159	0.156	0.159	0.157	0.166
5	0.149	0.164	0.172	0.170	0.159	0.162	0.159	0.157	0.152	0.165
6	0.168	0.165	0.174	0.155	0.157	0.159	0.158	0.164	0.145	0.156
7	0.157	0.160	0.158	0.159	0.159	0.168	0.160	0.171	0.159	0.161
8	0.165	0.161	0.150	0.164	0.158	0.157	0.159	0.145	0.165	0.155
9	0.156	0.166	0.180	0.159	0.167	0.164	0.160	0.146	0.152	0.154
10	0.170	0.163	0.156	0.154	0.152	0.157	0.147	0.148	0.158	0.152
11	0.162	0.156	0.160	0.157	0.151	0.154	0.174	0.176	0.158	0.161
12	0.152	0.155	0.158	0.147	0.157	0.164	0.160	0.148	0.158	0.147
13	0.153	0.151	0.156	0.161	0.164	0.165	0.152	0.158	0.154	0.149
14	0.159	0.149	0.151	0.159	0.147	0.160	0.155	0.153	0.158	0.157
15	0.157	0.173	0.167	0.169	0.163	0.155	0.152	0.154	0.162	0.175
16	0.174	0.160	0.162	0.164	0.149	0.161	8.155	0.161	0.165	0.158
17	0.164	0.157	0.152	0.171	0.159	0.164	0.174	0.159	0.167	0.155
18	0.164	0.154	0.154	0.164	0.154	0.154	0.158	0.157	0.160	0.156
19	0.172	0.168	0.174	0.160	0.159	0.161	0.161	0.172	0.159	0.155
20	0.170	0.159	0.154	0.159	0.153	0.161	0.165	0.157	0.163	0.160

SUBGROUP NUMBER	X-BAR CHART DATA 1	2	3	4	5	6	7	8	9	10
MEAN	0.1619	0.1600	0.1615	0.1603	0.1579	0.1599	0.1595	0.1595	0.1588	0.1580
MEAN MEAN	0.1597	0.1597	0.1597	0.1597	0.1597	0.1597	0.1597	0.1597	0.1597	0.1597
UPPER CONTROL LIMIT (UC	0.1643	0.1643	0.1643	0.1643	0.1643	0.1643	0.1643	0.1643	0.1643	0.1643
LOVER CONTROL LIMIT (LC	0.1551	0.1551	0.1551	0.1551	0.1551	0.1551	0.1551	0.1551	0.1551	0.1551

SUBGROUP NUMBER	RANGE CHART DATA 1	2	3	4	5	6	7	8	9	10
RANGE	0.0252	0.0245	0.0303	0.0236	0.0220	0.0144	0.0293	0.0386	0.0215	0.0281
MEAN RANGE	0.0257	0.0257	0.0257	0.0257	0.0257	0.0257	0.0257	0.0257	0.0257	0.0257
UPPER CONTROL LIMIT (UC	0.0407	0.0407	0.0407	0.0407	0.0407	0.0407	0.0407	0.0407	0.0407	0.0407
LOVER CONTROL LIMIT (LC	0.0107	0.0107	0.0107	0.0107	0.0107	0.0107	0.0107	0.0107	0.0107	0.0107

Figure 8.14 Example of data captured in spreadsheet

popular spreadsheet programs, including designing the graphs discussed below. At least one basic textbook for beginners wishing to apply SPC to extrusion lines is available [51].

Several graphical displays were designed into the spreadsheet, including $\overline{x}$ (average) and R (range) charts, histogram of the measured dimensions, and a normal curve generated from the standard deviation and average, as well as a figure showing a twenty point rolling average of the standard deviation and average over a "standard run".

Data presented in the graphical displays made it possible to instantly determine whether the process was in control and to instantly see what impact changes made to the extrusion line had on process capability. For this application and the problems observed, it was determined that measurements should be recorded once a second giving a continuous view of the process over almost 7 minutes, which was adequate for the analyses to be made.

8.6.4 The Extrusion Line

The extruder initially used was a 1968 2½ inch 24:1 NRM extruder running a general purpose screw and a Vari-Drive AC speed controller. It had four zone analog temperature control with digital temperature displays. No other extrusion parameters were measured.

Diehead and die temperatures were controlled with Variac transformers connected to the heater bands. Direct temperature measurements were not made. Instead, temperatures were controlled by adjusting the approximate output voltage indicated by the pointer on the knob on the Variac.

A standard breaker plate feeding an open cylindrical diehead with no streamlining backed a 40/20 square mesh twilled weave screen pack. A flat die plate with a minimum land length to part thickness of about 7:1 was bolted to the diehead. The die was designed for a 250% pull down.

The profile was pulled over a standard air table and through a conventional water bath using an old catapuller with mechanical speed control and no instrumentation to measure puller speed. No chillers were used and the water temperature was not monitored.

The success of this project was a result of the quality and attitude of the profile extrusion company and not its equipment. It should be noted that this profile extrusion company has been in business 33 years, is noted for its outstanding customer service, and has always been profitable. It has a large, loyal customer base, most of whom have been with it for 15 years or more.

8.6.5 Extrusion Line Changes, General

After the profile extrusion company conducted several preliminary test runs, the customer targeted four areas for modification in the extrusion line. These changes were made one at a time. The results presented in the following are based on a single profile.

Variables such as air table blower location, physical line guides, process temperatures, and run speeds were optimized after the changes discussed below were made.

Several data runs were made after each change to the production line. A standard data run for this project was defined as a collection of 400 measurements made at one second intervals. Each individual measurement was the average of twenty-four measurements made at 0.86 ms intervals. All figures and numerical data presented are based on a standard data run.

Many data runs were also made before and after each line change at different sampling and averaging rates and the differences in the results from these data runs were also analyzed. Analysis of these differences was useful and provided additional insight into some of the problems encountered. While further discussion would be relevant, space does not permit this digression.

The control charts that were generated for each data run showed that the process remained in control throughout all of the changes that were made to the line.

For brevity, only the data relevant to the process potential index (CP) for tolerance variation are discussed, since all tests showed that stability of the average was not an issue.

8.6.6 Extrusion Line Changes, Details

Figure 8.15 gives an overview of how standard deviation decreased as the following changes were made to the line and how production speeds were in increased.

1. The first change made was to replace the puller with a custom-designed puller built specifically for the customer's profiles. This puller had a digital speed control with absolute drive speed accuracy of ± 1.0 fpm with a maximum fluctuation of ± 0.5 fpm for speeds between 50 and 100 fpm.

   Installing a puller with known speed characteristics immediately eliminated one potential variable and made it easier to identify other problem variables. This change provided the single largest reduction in the standard deviation of the key dimension and brought the CP up from 0.35 to 0.61.

2. The second change was to use a newer extruder with a more current overhaul. This extruder was nearly identical to the first but had been built sometime around 1983. Changing to this extruder brought the second most dramatic reduction in standard deviation, bringing the CP to 0.91.

   After the profile was running on the new machine, standard deviation data were collected over a range of extruding speeds. While no numerical studies were made, it was clear that standard deviation increased more than proportionally with increasing screw speed.

   On older extruders with mechanical screw speed control, screw speed variation decreases as a percentage of screw speed as screw speed is increased. This suggested that the increase in standard deviation with extrusion speed came from variation in die flow, perhaps caused by melting or mixing deficiencies in the extruder.

3. A pull down of 250% is much larger than would normally be used for the profile being extruded. Under some circumstances, a large pull down masks melt fracture and eddy flow imperfections by stretching them out over a longer section of the extrusion, thus

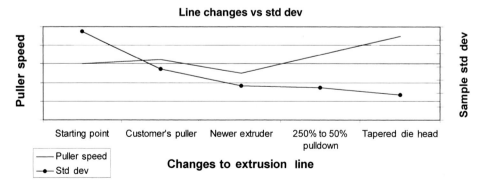

Figure 8.15 Changes in standard deviation resulting from changes to the extrusion line

causing them to be less apparent. A large pull down may also contribute to draw resonance.

The easiest way to start exploring these conjectures was to build a die with a much smaller pull down and evaluate the results. The customer suggested using a 25% pull down, but due to the extruder's past experience, a compromise pull down of 50% was used for the next die design.

The new design slightly reduced the rate at which the standard deviation increased with screw speed. It was possible to increase extrusion speed by 20 fpm while achieving a slight reduction in the standard deviation (increase in the CP) relative to the initial conditions.

4. This left streamlining the flow in the die as the next obvious change to make. This proved to be the most important change made to the die design. The CP increased from 0.96 to 1.25 while the rate of increase in standard deviation with increase in screw speed was greatly reduced. Further adjustments to the tooling and line setup have increased the CP to nearly 1.6 in actual production.

8.6.7 Conclusions

Simple SPC methods and inexpensive instrumentation and software can be used to substantially improve the performance of almost any extrusion line with limited instrumentation and without overall line control. Most low technology extrusion companies have the capability to make higher precision and higher quality parts than they presently make at no increased cost to themselves, but are not motivated to do so since their present customer base is satisfied.

Some profile extrusion companies fitting this description are receptive to trying new ideas. Extrusion companies are easier to convince if the customer is willing to accept the initial development risks and can show the profile extrusion company how success can deliver a meaningful increase in its sales and profitability.

One way to accomplish this is for the customer to take his own basic SPC instrumentation to the profile extrusion company and demonstrate its value and power during the development of the customer's profile.

9 Other Tools to Improve Process Control

9.1 Introduction

Thus far, we have concentrated on just one statistical tool to improve process control, the conventional SPC as originally developed by Shewhart. It should be realized, however, that there are other tools that can, and should, be used to achieve world-class quality. Keki Bhote [19], in his book *World Class Quality, Using Design of Experiments to Make it Happen,* lists the following characteristics of a world-class company:

Management	• Quality a superordinate value • Prevention a way of life
Organization	• Teams and focused factories • *All* employees responsible for quality
System	• Continuous, never-ending improvement • Intangible quality costs attacked
Tools	• Design of experiment • Quality function deployment • Multiple environment overstress test • Total productive maintenance
Customer	• Customer enthusiasm • Next operation as "customer" pervades the organization • Design for zero variation and zero failures • Multiple environment overstress test • User-friendly built-in diagnostics
Supplier	• Supplier an extension of the company • Quality, cost, and cycle time help to suppliers • Self-certified suppliers
Process	• Scrap eliminated • Inspection/testing greatly reduced • CpK > 5.0 • Factory overall efficiency > 85%
Support	• Internal customer evaluation replaces evaluation by boss • Financial incentives/penalties established within next operation as customer
People	• "Every employee a manager" • Self-directed work teams • Manager is a consultant, not boss • From management to leadership

It is interesting to note that under tools, the first item mentioned is design of experiments and *not* SPC! This has to do with the inherent limitations of conventional SPC. In the previous chapters we have explored the main elements of SPC and what we can do with SPC. What is equally important to know is what we *cannot* do with SPC. Using SPC we can assess and quantify the variations in the process. However, we cannot determine the exact causes of variation. SPC tells us to eliminate special causes of variation, but it does not tell us how. In addition, in many cases we also need to reduce or eliminate common causes of variation to achieve top quality production. Again, SPC does not give us much guidance here.

SPC is a useful tool once the process has been developed and regular production takes place. In the development stage, however, one of the most important tools is the design of experiments. Thus in a logical quality system, design of experiments, or DOE, is used before SPC is practiced. DOE is a development tool to arrive at a robust manufacturing process. Once the manufacturing process has been established, SPC is used to make sure that the process continues to behave well. In this sense, SPC is a maintenance tool, while DOE is a development tool.

The book by Lahey and Launsby [48] is devoted exclusively to design of experiments in injection molding; there is no such book for extrusion. This book presents several case studies using Taguchi, Box-Behnken, fractional factorial, and full factorial designs. Each case study presents a real-world situation encountered by injection molding companies.

9.2   Design of Experiments (DOE)

As discussed earlier, DOE is useful in developing a robust process without chronic quality problems. Another important application of DOE is in existing processes that exhibit chronic quality problems. Most likely, this will be a process where DOE was not used in the process development stage. The objectives in DOE are:

1. To identify the important variables that affect quality
2. To determine the main effects and the interaction effect of the important variables
3. To reduce the variation of the important variables
4. To open up tolerances on the unimportant variables to reduce cost

There are three main approaches to DOE: the classical, Taguchi, and Shainin. Much of the early development of the classical DOE is the work of R. A. Fisher [21] and F. A. Yates [22] in the 1930s at an agricultural experimental station in England. After World War II, their work has been extended to problems in the manufacturing industries by many statisticians. G. E. P. Box and his colleagues [20, 23] were instrumental in developing evolutionary operation, a technique for optimizing industrial processes.

The Japanese statistician Genichi Taguchi modified the classical DOE using the known technique of orthogonal arrays [14]. Taguchi did more than modify the classical DOE; he developed a complete system to develop specifications, to engineer the design to these specifications, and manufacture the product to specifications. The Taguchi method is used by a number of companies.

The third approach to DOE is the system of tools developed by Dorian Shainin. The Shainin method is the least known of the three approaches; this is ironic, since it is the most powerful and user-friendly. The Shainin method uses full factorial experimental designs. However, the full factorial experiment is preceded by a number of other steps, such as multi-vari charting, components search, paired comparison, and variables search. The steps preceding the full factorial are taken to reduce a large number of variables, 20 to 1000, to a small number, four or less, which makes full factorial experimentation practical and useful. Table 9.1 compares the three DOE approaches.

It is clear that the Shainin method has considerable advantages over the classical DOE and the Taguchi method. Since the latter two methods have received extensive coverage in the literature, while the former has received scant attention, we will elaborate on the Shainin methodology. One of the reasons that the Shainin method has not received the attention it deserves, is that, unfortunately, Shainin has not written the definitive book on the Shainin methods, only a collection of articles. The only people that have access to the details of the Shainin method are employees of companies that hire Shainin for his consulting services. One of the few books that deals with the Shainin method in some level of detail is the book by Bhote [19], an employee of Motorola who was exposed to Shainin's teachings for a considerable period of time.

9.3 The Shainin Methodology

Figure 9.1 shows the seven tools of the Shainin DOE method. The basic philosophy behind this approach is: "Let the parts do the talking." This means that it is important to critically examine the parts produced to try to learn what may be going wrong in the production process. Do not make guesses, but learn as much as possible from the physical evidence presented by the process. By an intelligent analysis of the parts, the number of possible causes of problems can be reduced drastically.

The Monte Carlo simulation can be used if the mathematical relationship between the input variables are known. Components search is used at the prototype stage if only two to six units are available. Multi-vari charts are used in engineering runs, pilot runs, or full production with thirty or more units available.

Improving a process is like detective work: clues have to be gathered with each DOE tool. With each tool, the number of possible culprits is further reduced until the main cause or variable is captured. Shainin calls the number one culprit cause the *Red X*; the second most

Table 9.1 Comparison of Three Approaches of DOE, after Bhote [19]

Characteristic	DOE approach		
	Classical	Taguchi	Shainin
Principle techniques	• Fractional factorials • EVOP*	• Orthogonal arrays	• Multi-vari charts • Paired comparison • Variables search • Full factorial
Effectiveness	• Good in the absence of interactions • Poor if interactions are present • Limited optimization	• Good in the absence of interactions • Very poor if interactions are present • Very limited optimization	• Extremely powerful regardless of interactions • Retrogression rare • Maximum optimization
Cost/time	• Moderate (8 to 50 experiments)	• Moderate in the absence of interactions • High if there are interactions	• Low (3 to 30 experiments)
Complexity	• Moderate • ANOVA required • 3 to 5 days of training	• High • Inner and outer arrays multiplied • S/N* and ANOVA* • 3 to 10 days of training	• Low • Mathematics are simple • 1 to 2 days of training
Statistical validity	• Low • Saturated designs with confounding of main and interaction effects	• Poor • Highly saturated designs with extreme confounding of main and interaction effects	• High • Clear separation of main and all low and high order interactions
Versatility	• Low • Only two tools available	• Poor • Only one tool available	• High • 20 tools available to tackle many problems
Scope	• Requires hardware • Main use in production	• Can be used at design stage • Main use in production	• Requires hardware • Can be used at prototype, preproduction, and production stages
Ease of implementation	• Moderate • Statistical knowledge and computers required	• Poor • Statistical knowledge and computers required • Engineers "turned off" by complexity and modest results	• High • Minimal statistical knowledge required • Engineers turned on by simplicity and excellent results

* EVOP = evolutionary operation; ANOVA = analysis of variance; S/N = signal-to-noise ratio.

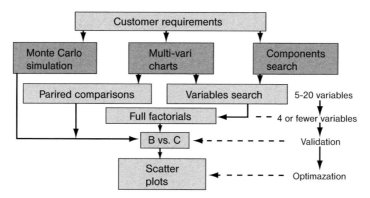

Figure 9.1 An overview of the Shainin tools for DOE

important cause is the *Pink X*, the third most important the *Pale Pink X*. When the *Red X*, *Pink X*, and the *Pale Pink X* have been identified, well over 80% of the variation allowed within the specification limits is eliminated. Thus, a CpK of 5.0 can be achieved with just one, two, or three DOE experiments.

9.3.1 Tools to Generate Clues

9.3.1.1 Multi-Vari Charts

Multi-vari (MV) charts were originally developed by L. A. Seder [24]. These charts are used to discover *causes* of variation. Note the difference with control charts, which are used to determine variation, not the causes of variation. In the MV chart, the process variation is divided into families. The chart is used to determine whether the major variation is positional, cyclical, or temporal. Positional variation can be variation within a single unit or in a batch, for example, the cavity-to-cavity variation in multicavity injection molding. Cyclical variation is variation between consecutive units drawn from the process, while temporal variations are variations from hour to hour, shift to shift, day to day, and so on.

An example of an MV chart is shown in Fig. 9.2. This chart shows cylindrical shafts drawn from a manufacturing process using an old turret lathe. The shafts have a diameter requirement of 0.0250 in ± 0.001 in or a tolerance of 0.002 in. The actual variation in the process was measured and it was found that the standard deviation of the diameter was 0.000417 in. With a standard deviation of 0.00417 in the 6σ process variation becomes 0.0025 in. The process capability in this case is CP = 0.002 in/0.0025 in = 0.8; this is quite a low value, leading to a substantial number of out-of-spec products.

In order to try to find the causes of variation, an MV chart was made, taking five sets of samples at 8, 9, 10, 11, and 12 a.m. Each sample consisted of three consecutive units from the process. Four readings were made on each shaft, two readings 90 degrees apart (x and y)

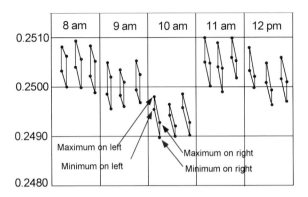

Figure 9.2 Example of a Multi-Vari chart for cylindrical shafts

on the left side and two readings 90 degrees apart on the right side of the shaft. The difference between two readings on the same side of the shaft are an indication of out-of-roundness, while the difference between left and right side readings are an indication of taper of the shaft.

It is easy to see from the figure that the greatest variation that occurs is temporal (from one hour to another), with the biggest change from 10 to 11 a.m. This provided a clue because of the coffee break between 10 and 11 a.m. It was postulated that the lathe cooled down during the coffee break, causing the large increase at 11 a.m. After looking into this further, it was found that the level of the cooling fluid in the tank was low. After filling the tank with cooling fluid to the correct level, the time-to-time variation largely disappeared. This simple change eliminated about 50% of the allowed variation.

Further analysis of the MV chart indicates that there is considerable variation within each unit, 30% due to out-of-roundness and 10% due to taper. The out-of-round condition was traced to a worn eccentric bearing guiding the chuck axis. Installing new bearings essentially eliminated the out-of-round condition. The taper showed a consistent pattern with the left side always larger than the right side. This was due to the cutting tool not being parallel to the shaft axis; a simple adjustment reduced the taper to almost zero.

With some simple changes the process variation was reduced from 0.0025 in to 0.0004 in with a correspnding increase in CP from 0.8 to 5.0, an improvement of more than six times! This example illustrates how useful the MV chart can be in not only showing the variation in the process, but at the same time yielding clues on the cause of variation. With these clues, improvements to the process can be made quickly and efficiently.

9.3.1.2 Components Search

Components search is a technique designed to reduce a very large number of possible causes to a very small number. It is particularly applicable in assembly operations, where

9.3 Other Tools to Improve Process Control

the performance of a unit is dependent on a number of individual components. For this technique to be applicable:

- The performance of a unit must be measurable and repeatable.
- It must be possible to disassemble and reassemble the unit without a significant change in performance.
- There must be at least two assemblies with clearly different output levels.

There are four stages in components search. The first stage, *the ballpark stage*, determines if the *Red X* and *Pink X* are trapped among the factors to be investigated. The second stage, *the elimination stage*, is to eliminate all the unimportant main causes as well as the associated interaction effect. The third stage, *the capping run stage*, verifies that the important causes selected in the second stage continue to be important when combined and that the unimportant causes are indeed unimportant. The fourth stage, *factorial analysis*, uses a full factorial matrix to quantify the main and interaction effect.

Components search involves ten steps:

1. Select the best and worst units from a production run.
2. Determine what performance characteristic to measure and measure both units.
3. Disassemble the good unit twice; reassemble and remeasure it each time. Do the same with the bad unit. The readings of the three good units must all rank better than all three bad units. The difference, D, between the median of the good and bad units must exceed the average range, d_a, of nonrepeatability within each of the good and bad units by a factor of at least 1.25. This factor is required to establish that there is a significant and repeatable difference between the good and bad units.
4. Rank the components of the units in descending order of perceived importance based on engineering knowledge.
5. Switch the top-ranked component, A, from the good unit to the bad unit, and component A from the bad unit to the good one. Measure both units.

6a. If there is no change, component A is unimportant, go to component B, step 7.

6b. If there is a partial change, A is not the only important variable. Also, interaction effects may be important; go to component B, step 7.

6c. If there is a complete reversal in performance, A is the part with the *Red X* characteristic; the search can end here.

7. Restore both components A to the original good and bad units to make sure that the original condition is repeated. Repeat steps 5 and 6 with the next most important component, B, then C, and so on.
8. Eventually, the *Red X* family involving two or more components will be indicated. Interaction effects can be found in step 10.

9. Once the important components have been identified, a capping run of these components banded together in the good and bad assemblies is necessary to verify their importance.
10. The last step is to set up a factorial matrix to determine the main and interaction effects.

9.3.1.3 Paired Comparisons

This technique is similar to components search; it is used when:

- Components cannot be disassembled or reassembled.
- There are several good and a few bad parts that can be paired.
- There is a measurable characteristic that distinguishes good from bad.

Paired comparisons can be used in assembly work or processes or in test equipment, where there are similar units. It is also useful in failure analysis. The method involves five steps:

1. Select one good and one bad unit; call this pair one.
2. Observe in detail to determine differences between the two units.
3. Select a second pair of good and bad units and observe and note differences.
4. Repeat this process with a third, fourth, and fifth pair, until the differences show a repeatable pattern.
5. Disregard differences that do not follow a repeatable pattern.

In most cases, by the fifth or sixth pair the consistent differences will have been reduced to just a few factors. By this time there will be a strong clue as to the major cause of variation.

9.3.1.4 Variables Search

After the multi-vari chart, the components search (if applicable), and the paired comparison (if applicable), the number of possible problem causes should be reduced to about five to fifteen. Further reduction is achieved with variables search. The objective of a variables search is to identify the Red X, Pink X, and any interacting variables; also, the variables search aims to separate important variables from unimportant ones.

In a full factorial DOE, each variable is tested with each level of every other variable. When we have five to fifteen variables at two or three levels, this can lead to a very large number of experiments. For instance, six variables at three levels will require 3^6 or 729 experiments, in most cases an impractical number of experiments. To overcome this problem, various shortcuts have been applied, such as the fractional factorial design and the orthogonal array. However, these shortcuts are not accurate when interaction effects play a role, as they often do.

The purpose of the variables search is to reduce the number of variables to a small enough number that a full factorial becomes practical. The variables search consists of four stages;

the first stage involves the identification of the important variables and an initial analysis of their importance. The steps in *Stage 1* are:

1. Make a list of the most important variables or factors in order of importance; the variables are represented by the letters A, B, C, and so on.
2. Select two levels for each factor; the high level is expected to give good results, the low level is the typical deviation from the high level in regular production.
3. One experiment is run with all the factors at their high level and one experiment with all factors at their low level. These experiments are repeated twice; thus we have three experiments at high level and three at low level.
4. Apply the $D:d_a > 1.25$ rule as applied in step 3 of the components search.
5. If all three high levels are better than the three low levels, then Stage 1 is over, provided the $D:d_a$ ratio is greater than 1.25. If all three low level results are better, change the heading of high to low and vice versa.
6. If the conditions of step 5 are not met, switch one pair of the most likely factors from high to low and vice versa.
7. If the repeatability, d_a, is still poor, this indicates that an important variable has been omitted from the list compiled in step 1. Determine additional variables that may have an important effect and recompile the list in step 1 and repeat the steps 2 through 7.

The second stage involves determination of which factors are unimportant. *Stage 2* consists of the following steps:

1. Run two A_L/R_H tests; this means factor A at the low level and the remaining factors at high level. Follow this by two A_H/R_L tests. Determine the control limits from the average range as follows:
control limits = median ± 2.776 $(d_a/1.81)$

    Possible results:

    a) If the results of both pairs of tests fall inside the control limits, factor A and all its associated interaction effects are unimportant.
    b) If there is a complete reversal, A is the main effect Red X. All other factors are unimportant and can be eliminated — the variables search has ended.
    c) If one or both tests show results outside the control limits, but not a complete reversal, factor A cannot be eliminated.

2. Repeat step 1 with factor B if the results are a) or c). If c) results, then factor B cannot be eliminated and must be considered with some other factor(s).
3. Repeat step 1 with factor C if the result in step 2 is a). If a) results, then C is unimportant and can be eliminated. If c) results, then factor C cannot be eliminated and C plus some other factor(s) must be considered.

Stage 3, the capping run, consists of the following steps:

1. If the factors A and B show a partial reversal with readings outside the control limits, run a $A_H B_H R_L$ and $A_L B_L R_H$ test to determine if R can be eliminated. The search is not yet complete if one or more results fall outside the control limits. Continue with the next single factors until another factor shows out-of-cntrol results.
2. At this point, run a three factor capping run.

It is possible that a four factor capping run may be possible. However, this is rather unusual in actual industrial applications.

Stage 4 consists of a factorial analysis to quantify the main effects and interaction effects of the most important factors.

Variables search reduces process variability more than the conventional DOE or the Taguchi method of DOE, and it gives more reliable results. It is a good tool to separate the important variables from the unimportant ones. By relaxing tolerances on the unimportant variables, the manufacturing cost can be reduced considerably. The method is simple enough that it can be performed by technicians and operators. The cost of experimentation with the variables search is three to ten times less compared to the classical DOE or the Taguchi DOE.

9.3.2 Factorial Designs

Various experimental designs are used, such as block designs, Latin square, nested designs, and so on. Factorial designs are particularly useful for experiments that involve several variables at two or more levels. The main factorial design is the full factorial; this is a design where all combinations of variables at every level are run. If we have N levels and K variables, the total number of variables is N^K. Thus for a two-level design ($N = 2$) with three variables ($K = 3$), the number of experiments will be eight ($2^3 = 8$).

In order to overcome residual error inherent in all experimentation, these eight experiments should be repeated (or replicated), to make the total number of experiments sixteen. In a full factorial we want to obtain information on the effect of the main variables, any possible interaction effects, and the variance. Also important in factorial design is that the experiments are conducted in random order. Random number generators or tables with random numbers are useful to randomize the experimental sequence.

When the number of experiments is too large to run all combinations, a blocked factorial design can be used or a fractional factorial. As mentioned earlier, the drawback of these designs is that some or all interaction effects are lost. A more efficient method is to reduce the number of variables to a small enough number that a full factorial design can be used. The Shainin method does this by using multi-vari charts, components search, paired comparison, and variables search.

Factorial designs are very useful because they provide the maximum amount of information from a minimum number of experiments. The most common experimental technique is to vary one variable at a time to study its effect; this is called the one-at-a-time approach. Even though this makes for a very simple experimental plan, this approach often does not work, particularly when interaction effects play a role — this is often the case in industrial settings.

9.3.2.1 Interaction

Interaction occurs when the influence of one variable changes with the level of another variable. We can take an example from the plasticating unit of the injection molding machine. If we plot the effect of screw speed on output at different levels of the discharge pressure, we obtain a plot showing several lines shifted a certain amount, this is shown in Fig. 9.3.

The different values of the pressure do not change the relationship between screw speed and output other than shifting the curve. This is an example of *no interaction* between the two variables screw speed and pressure. As another example we will look at the relationship between extruder output and channel depth at different values of the helix angle of the flight. This is graphically shown in Fig. 9.4.

The helix angle is represented by Greek character ϕ. In this figure it is clear that the relation between output and channel depth changes markedly when the value of the helix angle is changed — there is no longer a simple shift in the curve, the shape of the curve changes

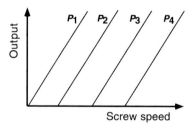

Figure 9.3 The effect of screw speed on output at different pressure levels

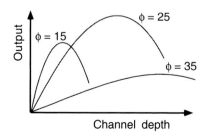

Figure 9.4 The effect of channel depth on output at different values of the helix angle

altogether! This is an example of *interaction* between channel depth and helix angle. In such a situation, one-at-a-time experimentation can lead to poor results. To illustrate this we can look at a three-dimensional plot of the output versus channel depth and helix angle; this plot can be determined from extrusion theory [12]. Figure 9.5 shows a three-dimensional contour plot of the output as a function of channel depth and helix angle.

In this figure the channel depth is shown along the horizontal axis, the helix angle in radians along the vertical axis; the curved lines in the figure are lines of constant output; the number at the curve indicates the output in pounds per hour. Different output regions, for instance from 100 to 150 lb/hr, are separated by the contour lines.

If we are trying to optimize the extruder screw for maximum output, we can run a trial where we run at three values of the channel depth, for instance 0.150, 0.200, and 0.250 in. The helix angle is kept constant in these trials at 0.2 radian. We will assume that we do not have the information shown in Fig. 9.5. From the trial at three channel depth values we find that the highest output occurs at $H = 0.250$ in, where the output is about 145 lb/hr.

The second step involves running a trial where we run at three values of the helix angle, for instance 0.2, 0.4, and 0.6 radians, while the channel depth is kept constant at 0.250 in. In this trial we find that the highest output is obtained at a helix angle of 0.4 radians, the output is about 205 lb/hr. From these one-at-a-time experiments we would conclude that the maximum output is 205 lb/hr and occurs at a depth of 0.25 in and a helix angle of 0.4 radians. It turns out that the true maximum is 240 lb/hr and occurs at a depth of 0.17 in and a helix angle of 0.6 radians.

This example shows how one-at-a-time experiments can lead us to the incorrect optimum, 205 lb/hr, compared to the true optimum of 240 lb/hr! Clearly, one-at-a-time experiments

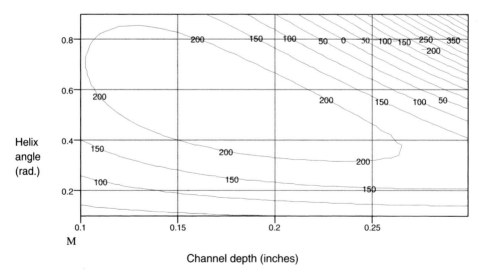

Figure 9.5 Contour plot of output as a function of channel depth and helix angle

are inappropriate when interaction effects are present. This is the case in many industrial problems.

9.3.2.2 Two-Level Factorial Design with K Factors

The 2^K factorial designs are often used in industrial applications. This design allows separate determination of the individual effects and the interaction effects of K factors. Different notations are used to identify the different trials; the most common notation is to show the high value as (+) and the low value as (−). The trials for a four-factor factorial are shown in Table 9.2. The actual trials are shown in the first four columns. The interactions are shown in columns five through fifteen.

In a 2^K factorial all K first-order effects can be estimated; further, all $K(K-1)/2!$ two factor interactions, all $K(K-1)(K-2)/3!$ three-factor interactions, and so on. Thus in a four-factor design we can estimate four first-order effects, six two-factor interactions, four three-factor interactions, and one four-factor interaction — this makes a total of 15 statistics. The magnitude and the sign of these statistics are not influenced by one another; the statistics are said to be orthogonal.

Table 9.2 Table of Signs for a 2^4 Factorial

A	B	C	D	AB	AC	AD	BC	BD	CD	ABC	ABD	ACD	BCD	ABCD
−	−	−	−	+	+	+	+	+	+	−	−	−	−	+
+	−	−	−	−	−	−	+	+	+	+	+	+	−	−
−	+	−	−	−	+	+	−	−	+	+	+	−	+	−
+	+	−	−	+	−	−	−	−	+	−	−	+	+	+
−	−	+	−	+	−	+	−	+	−	+	−	+	+	−
+	−	+	−	−	+	−	−	+	−	−	+	−	+	+
−	+	+	−	−	−	+	+	−	−	−	+	+	−	+
+	+	+	−	+	+	−	+	−	−	+	−	−	−	−
−	−	−	+	+	+	−	+	−	−	−	+	+	+	−
+	−	−	+	−	−	+	+	−	−	+	−	−	+	+
−	+	−	+	−	+	−	−	+	−	+	−	+	−	+
+	+	−	+	+	−	+	−	+	−	−	+	−	−	−
−	−	+	+	+	−	−	−	−	+	+	+	−	−	+
+	−	+	+	−	+	+	−	−	+	−	−	+	−	−
−	+	+	+	−	−	−	+	+	+	−	−	−	+	−
+	+	+	+	+	+	+	+	+	+	+	+	+	+	+

The sign of each interaction is found by multiplying the + and – signs of the components of that interaction. For example, the *AB* interaction is found by multiplying the sign of factor *A* with the sign of factor *B*. The estimated interaction is determined by adding all the observations carrying the + sign and taking their average, then all the observations carrying the – sign are added and averaged. The estimated interaction is then the difference between these two values. For instance, the *AB* interaction is:

$$(x_1 + x_4 + x_5 + x_8 + x_9 + x_{12} + x_{13} + x_{16})/8 - (x_2 + x_3 + x_6 + x_{10} + x_{11} + x_{14} + x_{15})/8$$

The *ABC* interaction is:

$$(x_2 + x_3 + x_5 + x_8 + x_{10} + x_{11} + x_{13} + x_{16})/8 - (x_1 + x_4 + x_6 + x_7 + x_{12} + x_{14} + x_{15})/8$$

In shorthand notation the interaction is calculated by $\bar{x}_+ - \bar{x}_-$ with the first term representing the average of the plus-values and the second term the average of the minus-values. There is a faster method to compute the interactions; this is the Yates' algorithm [22]. An important question in the analysis of DOE data is whether the observed difference between the (+) and (–) levels is due to sampling chance or is it more than chance can explain. In other words, how important is it that you do not reach the wrong conclusion? This is expressed as a probability number called statistical risk, for instance 0.05. With a risk of 0.05, the confidence that the conclusion is correct will be 1.00–0.05 = 0.95 or 95%. The statistical risk has to be selected before tests are run, because it will affect the number of tests necessary to obtain the desired information.

9.3.3 The *B* Versus *C* Analysis

In this analysis we compare two processes, *C* is usually the current process and *B* is the process that is thought to be better. However, both *B* and *C* can be new processes. *B* versus *C* is a nonparametric comparative experimentation; thus there are no measurements as in variables data, but only a ranking of units from best to worst. The advantage of this technique is that only small sample sizes are required to determine with a high degree of confidence that one is better than the other.

The conventional technique to compare two processes is to run process capability studies on both processes. This will typically require fifty to one hundred readings from each process. From the process capability study we will find the frequency distribution for each process. Four possible scenarios can occur, as shown in Fig. 9.6.

If the distributions correspond to case (1), there is obviously no difference between *B* and *C*; this is called the null hypothesis. In case (2), *B* is better in most cases but there is some overlap; thus some *B* units are worse than some *C* units. This situation is called the Pink *X* condition. In case (3), the worst units of *B* are as good as or better than the best *C* units; this is called the Red *X* condition. Finally, in case (4) all *B* units are better than any *C* unit; this is the super Red *X* condition.

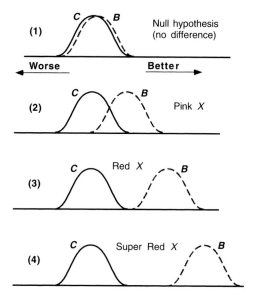

Figure 9.6 Four distributions of B and C processes

The problem with the conventional approach is that it requires a minimum of fifty readings from each process. With the B versus C test we typically use only three B units and three C units. The theory of B versus C is based on the formula for combinations. If there are two B units and two C units, they can be arranged in six possible ways, as shown in Table 9.3.

There are only six ways in which the two Bs and two Cs can be ranked in any order. Only one out of these six, column 1, shows the two Bs on top and the two Cs at the bottom. In this case, the null hypothesis is rejected. When we cannot reject that outright, the distributions shown in cases (3) and (4) of Fig. 9.6 are not possible. There is a one in six, or 16.7%, chance that with two Bs on top and two Cs at the bottom, this is through chance alone. With three B units and three C units the number of ways in which they can be arranged is shown in Table 9.4.

There are twenty ways in which the Bs and Cs can be arranged. There is only a chance of one out of twenty that all three Bs rank above the three Cs. In this case, concluding that B is

Table 9.3 Combinations with Two B Units and Two C Units

	1	2	3	4	5	6
Best	B	B	B	C	C	C
↑	B	C	C	B	B	C
↓	C	B	C	B	C	B
Worst	C	C	B	C	B	B

Table 9.4 Combinations with Three B Units and Three C Units

	1	2	3	4	5	6	7	8	9	10	11	12	13	14	15	16	17	18	19	20
Best	B	B	B	B	B	B	B	B	B	B	C	C	C	C	C	C	C	C	C	C
↑	B	B	B	C	C	C	C	C	C	C	B	B	B	B	B	B	C	C	C	C
	B	C	C	B	C	B	B	C	C	C	B	B	B	C	C	C	B	B	B	C
	C	B	C	B	B	C	C	B	B	C	B	C	C	B	B	C	B	B	C	B
↓	C	C	C	C	B	B	C	B	C	B	C	B	C	B	C	B	B	C	B	B
Worst	C	C	B	C	C	C	B	C	B	B	C	C	B	C	B	B	C	B	B	B

better than C carries only a 5% risk — this is called the α (alpha) risk. If it is concluded that B is better than C, while in reality B is no better than C, the experimenter will have committed a type I error. The probability of committing such an error is the alpha risk.

If the experimenter decides that B is the same as C, while actually B is better, the experimenter has committed a type II error; its probability is the β (beta) risk. Thus the alpha risk is the risk of rejecting the null hypothesis when no improvement exists. The beta risk is the risk of accepting the null hypothesis when, in fact, an improvement exists. It is necessary to randomize the sequence of testing in B versus C comparison, just as it is in factorial experiments. In production it is often tempting to run three Cs with the existing process and then change over to the B process. This is not permissible, however, since that would defeat randomization and the results could lose their validity.

The steps involved in B versus C are the following:

Step 1: Select an acceptable level of alpha risk.

Step 2: Decide on sample sizes for B and C tests.

Step 3: Randomize and conduct the tests.

Step 4: Rank order of results.

Step 5: Apply decision rule.

Step 6: Separate the means: the beta risk.

The sample size is determined by the alpha risk as shown in Table 9.5.

The k-factor in columns 4, 5, and 6 indicates the distance between the means ($\mu_B - \mu_C$), where k is given by: $k = (\mu_B - \mu_C / \sigma)$. In Table 9.5 it is assumed that the standard deviations (σ or sigma) for the two processes are the same ($\sigma_B - \sigma_C$). If we select an alpha risk of 0.05 and take a sample size of three Bs and three Cs, then the beta risk at $k = 1.6$ is 0.50. This means that if the mean of B (μ_B) is 1.6σ above the mean of C (μ_B), the difference will be detected only 50% of the time. If $k = 2.9$, the beta risk is 0.10; in this case, the difference will be detected 90% of the time. If $k = 3.3$, the beta risk is 0.05 and the difference will be detected 95% of the time. This is illustrated in Fig. 9.7.

Table 9.5 B versus C for One-Tailed Test (i.e., B Better than C, Not Just Different)

Alpha risk	Number Bs	Number Cs	k (β = 0.50)	k (β = 0.10)	k (β = 0.05)
0.001	2	43	3.0	4.0	4.3
	3	16	2.5	3.6	3.9
	4	10	2.3	3.4	3.8
	5	8	2.2	3.4	3.7
	6	6	2.2	3.3	3.7
0.01	2	13	2.3	3.4	3.8
	3	7	2.1	3.2	3.6
	4	5	2.0	3.1	3.5
	5	4	2.0	3.1	3.5
0.05	1	19	2.5	3.6	3.9
	2	5	1.7	3.0	3.4
	3	3	1.6	2.9	3.3
	4	3	1.7	3.0	3.4
0.10	1	9	2.1	3.2	3.6
	2	3	1.4	2.7	3.2
	3	2	1.4	2.7	3.2

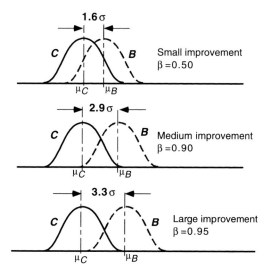

Figure 9.7 Three Bs and Cs: alpha risk = 0.05

B versus C can be used not only in manufacturing processes, but also in marketing, sales, advertising, and so on. Some of the applications of B versus C are:

- Evaluation of design changes, process changes, and material changes
- Evaluation of customer preferences
- Reliability evaluations
- Evaluation of sales/marketing/service practices
- Evaluation of human relations policies
- Evaluations of promotions/advertising

9.3.4 Scatter Plots to Determine Appropriate Tolerances

Tolerances are frequently set by a haphazard process, such as boilerplate figures, suppliers catalog recommendations, and so on. There is, however, a very effective method of setting tolerances using a graphical method called scatter plots. The scatter plots method is used to fine-tune the level of a critical variable to determine its best level, *the design center,* and the realistic tolerances. Scatter diagrams were discussed earlier in Section 5.6.

Scatter plots used to be called realistic tolerance parallelogram plots. They are used to relate the target and tolerance values for a few important input variables (X) to the output quality characteristic requirement (Y). These plots are not used for problem solving, because they illustrate association and not necessarily cause and effect. Scatter plots are a substitute for more sophisticated techniques such as evolutionary operation (EVOP). The advantages of scatter plots are their simplicity and their graphical, nonmathematical approach.

Figure 9.8 shows three independent input variables, X_1, X_2, and X_3, affecting a dependent output Y.

The question with Fig. 9.8 is, which variable is the Red X? One might think that the variable with the greatest slope is the Red X (in this case X_3); however, this is not true because the

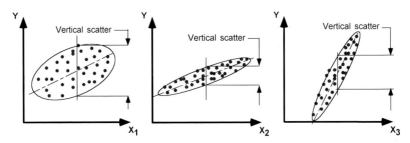

Figure 9.8   Interpretation of vertical scatter

horizontal scale is arbitrary. The Red X is the variable that has the smallest vertical scatter, thus variable X_2 in Fig. 9.8. The vertical scatter is a measure of the contribution of all other input variables (Xs) other than the one being plotted. Therefore, the Red X will have the smallest vertical scatter, because the Red X itself has the largest effect on output Y.

The procedure for determining realistic tolerances is as follows:

1. Select the output characteristic of interest, Y.
2. Select a range of values for the Red X variable that is likely to achieve optimum Y. Run thirty values of the Red X and determine the corresponding Y values and plot the results.
3. Draw the median line through the thirty points; this line is called the line of regression. Draw two lines parallel to the line of regression such that all but one and a half of the thirty points are contained between the two parallel lines. These are the scatter boundary lines. The vertical intercept through the boundary lines is the variation in Y due to all variables other than the one being plotted.
4. If Y is a customer requirement in terms of specification limits, plot these on the Y axis and draw two lines from them parallel to the X axis to the point where they intersect with the scatter boundary lines; see Fig. 9.9. Draw lines, parallel to the Y axis, from the intersecting points down to the X axis.
5. The horizontal intercept on the X axis determines the maximum tolerance permitted. This would correspond to a CP of 1.0. Using the concepts of precontrol (see Section 9.4), the horizontal intercept can be divided into four equal parts and only the middle half is allowed as the preferred tolerance for the Red X. This will ensure a CP of 2.0.
6. These target values and tolerances should be compared to existing values and tolerances and changes should be made to ensure zero defects.

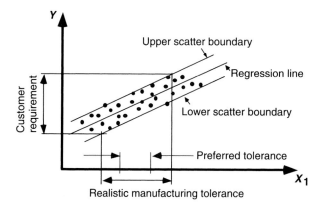

Figure 9.9 Tolerance parallelogram to determine realistic and preferred tolerances

9.4 Precontrol

Conventional statistical process control is strongly associated with the control chart or Shewhart chart, developed around 1930. The techniques used in conventional SPC have been described in detail in Chapters 4 through 8. Even though conventional SPC is quite useful, it has some important limitations. Control charts are useful to detect and quantify variation. However, control charts are not useful in finding and eliminating the causes of variation. Control charts are a good tool to analyze the symptoms, but not the root causes of variation. The main use of control charts is to maintain a process under control, once its variation has been reduced by the techniques described in Sections 9.1 through 9.3.

Precontrol was developed in the 1950s by a group of statisticians at the consulting company of Rath and Strong. This team, consisting of Warren Purcell, Frank Satterthwaite, Bill Carter, and Dorian Shainin, was given the task to see if the $\bar{x}$ and R method could be modified to work with smaller lots. The precontrol method was developed as a result of this effort; the theoretical basis of the method was developed by Frank Satterthwaite. The mechanics of precontrol are surprisingly simple and consist of four steps:

Step 1: Divide the tolerance by four. The boundaries of the middle half of the tolerance become the precontrol (P-C) lines; the zone between the P-C lines is called the green zone. The two zones between the P-C lines and the specification limits are called the yellow zones. The two zones beyond the specification limits are called the red zones. This is illustrated in Fig. 9.10.

Step 2: To determine whether the process is capable, five consecutive units have to be taken from the process. If all five fall within the green zone, the process can be considered in control and production can begin. In this case, the process capability will be at least 1.33. If not all five units fall within the green zone, the process has to be improved to reduce variation.

Step 3: When production has started, take two consecutive units from the process at regular intervals. The following situations can occur:

1. Both units fall within the green zone: continue production.
2. One unit in the green zone and one in the yellow zone: continue production.
3. Both units fall in the same yellow zone: adjust the process.
4. The two units fall in different yellow zones: stop process and reduce variation.
5. One unit falls into the red zone: stop process and fix the problem(s).

When the process is stopped, as happens in situation 3, 4, or 5, and the cause of variation has been found and reduced, step 2 has to be taken again before production can start again.

Step 4: The time between taking two consecutive units from the process is determined by dividing the average time period between two stoppages by six. For instance, if the average time between two stoppages is one and a half days (36 hours), the time between two consecutive units will be 6 hours.

9.4 Other Tools to Improve Process Control

The curve that is shown in Fig. 9.10 is a special case for a process with the process width equal to the tolerance (CP = 1.0). Clearly, the color zones are chosen to correspond to the colors of a traffic light, where red means stop, yellow caution, and green means keep going. The theory of precontrol is based on the multiplication theorem and the binomial distribution. The following are some of the statistical characteristics of precontrol.

- The worst alpha risk (stopping when not necessary) is about 2.0%.
- The worst beta risk (not stopping when necessary) is about 1.5%.
- When CPK < 0.8, precontrol will stop the process ninety-nine times out of a hundred.
- When CPK > 1.33, precontrol becomes most productive.
- When CPK > 2.0, precontrol will allow thousands of units to be produced without a single reject.

The precontrol system uses the carrot and the stick approach. It penalizes poor quality by shutting down the process frequently — this will force people to improve the process. On the other hand, it rewards good quality by requiring less and less frequent sampling. It is not necessary that the data from the samples be plotted on a chart as long as the data are recorded. However, if graphical data are required, the data can be plotted, and thus a precontrol chart can be created. From the sample data the CP and CPK can be calculated and the frequency distribution can be determined.

Precontrol can be used when there is a one-sided specification limit. In this case, we can use a single precontrol line located midway between the target and the specification limit. Precontrol can also be used for attributes data by converting an attribute into an artificial variable, using a scale from 1 to 10. The number 10 corresponds to a perfect quality characteristic, while number 1 corresponds to the worst possible quality characteristic. Other scales can be used as well, for instance from 1 to 5.

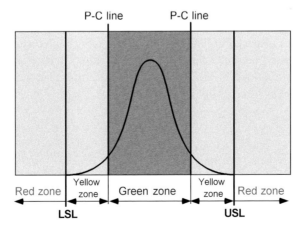

Figure 9.10 The three color zones in precontrol

One of the major advantages of precontrol is that it requires far less sampling than control charts. This can be important in injection molding, for instance if we are analyzing variation in a thirty-six-cavity mold. To construct a control chart requires a minimum of fifty measurements; thus we would need to make about 3600 measurements to chart all fifty cavities. With precontrol we need only five measurements for each cavity; thus we would need to make only a hundred eighty measurements. Once process capability is confirmed on each parameter, only the largest and smallest cavity need to be monitored.

9.4.1 Comparing Precontrol to Control Charts

It is clear from the information above that precontrol is much easier to use than control charts. Table 9.6 compares some important characteristics of precontrol to conventional control charts. From the table, it is quite clear that the precontrol in many, if not most, respects is superior to the control chart method.

The obvious question can then be asked: "If precontrol is easier and better, why are most companies still using control charts?" There are probably two answers to this question. One, many people are not aware of the precontrol method. Most books, classes, and seminars on SPC cover only the conventional SPC methods, without covering precontrol at all. Two, once conventional SPC has become established in an organization, it is difficult to get an organization to change. This is the common resistance to change; the larger the organization, the more difficult it is to get the organization to change.

It is hoped that, based on this discussion of control charts and precontrol, companies not currently using precontrol will take a serious look at it. Certainly, for companies not presently practicing SPC, precontrol will allow an easy transition into SPC with a minimum of training and cost. Unfortunately, some large corporations that require their suppliers to use SPC do not recognize precontrol as an acceptable SPC method. This rather shortsighted policy discourages some companies from using precontrol, not only to their own detriment, but also to the detriment of their customers. Precontrol is a simpler, more effective method of process control. As a result, it will reduce the cost of quality. This cost reduction can be passed on to the customer, at least partially. If a customer does not allow its supplier to use precontrol, obviously, the customer cannot reap the benefits associated with a more effective process control method.

Table 9.6 Comparison of Control Charts to Precontrol

Characteristic	Control charts	Precontrol
Purpose	To identify the amount of variation caused by random and special causes	To prevent the manufacture of defects
Simplicity	Complex — determination of control limits is tedious	Simple – precontrol is the middle half of tolerance
Use by operators	Difficult — charts have to be constructed and interpreted	Easy — green and yellow zones
Training	At least several days up to several weeks	Precontrol can be taught in about 10 minutes
Mathematical complexity	Involved — averages, ranges and control limits have to be calculated (intimidating)	Minimal — only requires checking whether data falls in green, yellow, or red zone (unintimidating)
Small production runs	Not useful for production runs below 500 units	Can be used for production runs above 20 units
Recalibration of control limits	Frequent	None
Machine adjustment	Time-consuming — an adjustment requires a trial run of 80 to 150 units	Quick — only two units needed to determine whether adjustment is necessary
Frequency of sampling	Vague — no clear rules to set sampling frequency	Simple, clear rule — six samplings between two consecutive stoppages
Discriminating power	Weak, both α and β risks are high	Good — α risk < 2% while the β risk < 1.5%
Attribute charts	The p charts and c charts do not distinguish between defect mode type or importance	Attribute charts can be converted to precontrol charts by weighting defect modes and an arbitrary rating scale
Economy	Expensive – substantial training, many tedious calculations, paperwork, large samples, etc.	Inexpensive — little training, simple calculations, minimum paperwork, small samples

References

1. W. Shewhart, *Economic Control of Quality of Product,* Van Nostrand Reinhold, New York (1931)
2. K. Ishikawa, *Guide to Quality Control,* Nordica International Ltd., Hong Kong (1976)
3. A. J. Duncan, *Quality Control and Industrial Statistics,* 5th ed., Richard D. Irwin, Inc., Homewood, IL (1986)
4. H H. Hun et al., *The Measurement Process,* National Institute of Standards and Technology, Special Publication Number 300, Government Printing Office, Washington (1969)
5. Statistical Quality Control Handbook, Western Electric Group., Inc., Newark, NJ (1956)
6. G. Menges and H. Recker, Eds., *Automatisierung in der Kunststoffverarbeitung,* Hanser Publishers, Munich/Cincinnati (1986)
7. A. N. Sahagen, "Silicon-on-Sapphire Pressure Transducer," *Measurement & Control,* October (1987)
8. H. Görman and H. Pütz, "Thermocouples with Increased Accuracy for Plastics Processing Machines," *Industrial and Production Engineering,* January (1980)
9. H. E. Harris, "A Look at the Thinking Behind a New Extruder Temperature Control Approach," *Plastics Technology,* February, p. 22–29 (1982)
10. S. S. Wilks, *Mathematical Statistics,* John Wiley & Sons, New York (1962)
11. C. Goldsberry, "Molders Say Resin Inconsistency Remains a Big Quality Problem," *Plastics News,* March 4, p. 1 and 16 (1991)
12. C. Rauwendaal, *Polymer Extrusion,* 4th revised ed., Hanser Publishers, Munich/ Cincinnati (2001)
13. G. E. P. Box and N. R. Draper, "Isn't my Process Too Variable for EVOP?" *Technometrics,* Volume 10, p. 439–444 (1969)
14. G. Taguchi, *Introduction to Quality Engineering,* Asian Productivity Organization, Tokyo (1986)
15. J. A. Clements, "Process Capability for Non-Normal Distributions," *Qualityware* (1990)
16. R. L. Plackett and J. P. Burman, "The Design of Optimum Multifactorial Experiments," *Biometrika,* Volume 33, p. 305 (1946)
17. ASQC Automotive Division/AIAG (Automotive Industry Action Group), *Measurement Systems Analysis Reference Manual,* AIAG (1990)
18. J. M. Juran and F. M. Gryna, Eds., *Juran's Quality Control and Handbook,* 4th ed., McGraw-Hill, New York (1988)
19. K. R. Bhote, *World Class Quality, Using Design of Experiments to Make It Happen,* AMACOM, New York, (1991)
20. G.E.P. Box, W. G. Hunter and J. S. Hunter, *Statistics for Experimenters,* John Wiley and Sons, New York (1978)

21. R. A. Fisher, "The Theory of Confounding in Factorial Experiments, in Relation to the Theory of Groups," *Ann. Eugen.* Volume 11 (1942)

22. F. Yates, "The Design and Analysis of Factorial Experiments," *Imper. Bur. Soil Sci. Tech. Comm.* Volume 35 (1937)

23. G.E.P. Box and J. S. Hunter, "Condensed Calculations for Evolutionary Operation Programs," *Technometrics* Volume 1, p. 77–95 (1959)

24. L. A. Seder, "Diagnosis with Diagrams — Parts I and III," *Industrial Quality Control,* January, March (1950)

25. D. J. Wheeler and D. S. Chambers, *Understanding Statistical Process Control,* SPC Press, Knoxville, TN (1992)

26. E. L. Grant and R. S. Leavenworth, *Statistical Process Control,* 5th ed., McGraw-Hill, New York (1980)

27. J. van der Veen and P. Holst, *Median/Individual Measurements Control Charting and Analysis for Family Processes,* Northwest Analytical, Inc. (1993)

28. H. Bajaria and F. Skog, "Realistic Multistream Process Charting," *Quality,* December (1994)

29. J. Ogando, "CQI Software Goes Beyond SPC," *Plastics Technology;* February, p. 17–20 (1995)

30. D. Hunkar, "A Practical Approach to the Determination of Process Capability in Injection Molding," SPE ANTEC, Detroit, MI, p. 2186–2191 (1992)

31. J. Ogando, "Portable Analyzers Find Out What Ails Your Process," *Plastics Technology,* February, p. 54–63 (1995)

32. C. J. Rauwendaal, "Screw Design for Reciprocating Extruders," 49th SPE ANTEC, Montreal, Quebec, p. 132–136 (1991)

33. N. N., "Injection Molding Monitoring Systems," *Plastics Machinery and Equipment,* February, p. 20–21 (1994)

34. D. Schmock, "Analyzer Tests Machine on a PC," *Plastics World,* August, p. 15 (1996)

35. D. J. Wheeler, "Problems with Gauge R&R Studies," Paper presented at the 46th Annual Quality Congress of the ASQC, Nashville, TN (1992)

36. G. Menges and P. Mohren, *How to Make Injection Molds* 3rd ed., Hanser Gardner Publications, Cincinnati, OH (2001)

37. K. Stoeckhert, Ed. *Mold-Making Handbook for the Plastics Engineer,* Hanser Gardner Publications, Cincinnati, OH (1983). An updated edition of this book was more recently edited by G. Mennig (1999)

38. P. Kennedy, *Flow Analysis of Injection Molds,* Hanser Gardner Publications, Cincinnati, OH (1995)

39. S. S. Shapiro and M. B. Wilk, "An Analysis of Variance Test for Normality (Complete Samples)," *Biometrika,* Volume 52, p. 591–611 (1965)

40. C. Rauwendaal and K. Cantor, "Obtaining Melt Flow Properties Directly from an Extruder," SPE ANTEC, Orlando, FL (2000)

41. C. Rauwendaal and P. Gramann, "Non-Return Valve with Mixing Capability," SPE ANTEC, Orlando, FL (2000)

42. C. Rauwendaal, *Polymer Extrusion,* Hanser Gardner Publications, Cincinnati, OH (1994)

43. C. Rauwendaal, *Polymer Mixing, A Self-Study Guide,* Hanser Gardner Publications, Cincinnati, OH (1998)

44. C. Rauwendaal, "Design of Dispersive Mixing Sections," *International Polymer Processing,* 13, p. 28–34 (1999)

45. C. Rauwendaal, "Comparison of Two Melting Models," *Advances in Polymer Technology,* Volume 15, No. 2, p. 135–144 (1996)

46. Z. Tadmor and C. Gogos, *Principles of Polymer Processing* 2nd ed., Wiley-Interscience, New York, (2006)

47. B. Maddock, SPE ANTEC, New York, (1959); also *Soc. Plastics Engrs. J.,* Volume 15, p. 383 (1959)

48. J. Lahey and R. Launsby, *Experimental Design for Injection Molding,* Launsby Consulting (1998)

49. D. Hadden and C. Rauwendaal, "Reduce Costs and Get Better Product Using your Own SPC System on your Line at your Profile Extruder's Plant: A Customer's Perspective," SPE ANTEC, New York, p. 2973–2979 (1999)

50. S. Basu, F. Fernandez, and C. Rauwendaal, "A Simple Test for Sagging Behavior in Blow Molding," SPE ANTEC, San Francisco, CA, p. 723–726 (1982)

51. C. Rauwendaal, *Statistical Process Control in Extrusion,* Hanser Gardner Publications, Cincinnati, OH (1993)

Appendix I List of Polymer Acronyms

Plastic Acronyms

In the plastics industry it is common to define a polymer by the chemical family it belongs to, and assign an abbreviation based on the chemistry. However, many times instead of using the standardized descriptive symbol, often engineers use the tradename given by the resin supplier.

This book uses the standardized notation presented in Table I.1. The symbols which have been marked with an asterisk (*) have been designated by the ISO standards, in conjunction with the material data bank CAMPUS. Furthermore, the acronyms presented in Table I.1 may have additional symbols separated with a hyphen, such PE-LD for low density polyethylene, or PVC-P for plasticized PVC. The symbols for the most common characteristics are presented in Table I.2.

Table I.1 Alphabetical Overview of Commonly Used Acronyms for Plastics

Acronym	Chemical notation
ABS*	Acrylonitrile-butadiene-styrene
ACM	Acrylate rubber, (AEM, ANM)
ACS	Acrylonitrile-chlorinated polyethylene-styrene
AECM	Acrylic ester-ethylene rubber
AEM	Acrylate ethylene polymethylene rubber
AES	Acrylonitrile ethylene propylene diene styrene
AFMU	Nitroso rubber
AMMA	Acrylonitrile methylmethacrylate
ANBA	Acrylonitrile butadiene acrylate
ANMA	Acrylonitrile methacrylate
APE-CS	see ACS
ASA*	Acrylonitile styrene acrylic ester
AU	Polyesterurethane rubber
BIIR	Bromobutyl rubber
BR	Butadiene rubber
CA	Cellulose acetate

Continued on next page

Acronym	Chemical notation
CAB	Cellulose acetobutyrate
CAP	Cellulose acetopropionate
CF	Cresol formaldehyde
CH	Hydratisierte cellulose, Zellglas
CIIR	Chloro butyl rubber
CM	Chlorinated polyethylene rubber
CMC	Carboxymethylcellulose
CN	Cellulose nitrate, Celluloid
CO	Epichlorhydrine rubber
COC*	Cyclopolyolene-Copolymers
COP	COC-Copolymer
CP	Cellulose propionate
CR	Chloroprene rubber
CSF	Casein formaldehyde, articial horn
CSM	Chlorosulfonated polyethylene rubber
CTA	Cellulose triacetate
DPC	Diphenylene polycarbonate
E/P*	Ethylene-propylene
EAM	Ethylene vinylacetate rubber
EAMA	Ethylene acrylic acid ester-maleic acid anhydride-copoly
EB	Ethylene butene
EBA	Ethylene butylacrylate
EC	Ethylcellulose
ECB	Ethylene copolymer bitumen-blend
ECO	Epichlorohydrine rubber
ECTFE	Ethylene chlorotrifluoroethylene
EEAK	Ethylene ethylacrylate copolymer
EIM	Ionomer Copolymer
EMA	Ethylene methacrylic acid ester copolymer
EP*	Epoxy Resin
EP(D)M	see EPDM
EPDM	Ethylene propylene diene rubber
EPM	Ethylene propylene rubber
ET	Polyethylene oxide tetrasulfide rubber

Continued on next page

Acronym	Chemical notation
ETER	Epichlorohydrin ethylene oxid rubber (terpolymer)
ETFE	Ethylene tetrafluoroethylene copolymer
EU	Polyetherurethane rubber
EVAC*	Ethylene vinylacetate
EVAL	Ethylene vinylalcohol, old acronym EVOH
FA	Furfurylalcohol resin
FEP	Polyfluoroethylene propylene
FF	Furan formaldehyde
FFKM	Perfluoro rubber
FKM	Fluoro rubber
FPM	Propylene tetrafluoroethylene rubber
FZ	Phosphazene rubber with fluoroalkyl- or fluoroxyalkyl group
HIIR	Halogenated butyl rubber
HNBR	Hydrated NBR rubber
ICP	Intrinsically conductive polymers
IIR	Butyl rubber (CIIR, BIIR)
IR	Isoprene rubber
IRS	Styrene isoprene rubber
LCP*	Liquid crystal polymer
LSR	Liquid silicone rubber
MABS*	Methylmethacrylate acrylonitrile butadiene styrene
MBS*	Methacrylate butadiene styrene
MC	Methylcellulose (cellulose derivate)
MF*	Melamine formaldehyde
MFA	Tetrafluoroethylene perfluoromethyl vinyl ether copolymer
MFQ	Methylfluoro silicone rubber
MMAEML	Methylmethacrylate-exo-methylene lactone
MPF*	Melamine phenolic formaldehyde
MPQ	Methylphenylene silicone rubber
MQ	Polydimethylsilicone rubber
MS	see PMS
MUF	Melamine urea formaldehyde
MUPF	Melamine urea phenolic formaldehyde
MVFQ	Fluoro silicone rubber

Continued on next page

Acronym	Chemical notation
NBR	Acrylonitrile butadiene rubber
NCR	Acrylonitrile chloroprene rubber
NR	Natural rubber
PA	Polyamide (other notations see Section 6.7)
PA11*	Polyamide from aminoundecanoic acid
PA12*	Polyamide from dodecanoic acid
PA46*	Polyamide from polytetramethylene adipic acid
PA6*	Polyamide from e-caprolactam
PA610*	Polyamide from hexamethylene diamine sebatic acid
PA612*	Polyamide from hexamethylene diamine dodecanoic acid
PA66*	Polyamide from hexamethylene diamine adipic acid
PA69*	Polyamide from hexamethylene diamine acelaic acid
PAA	Polyacrylic acid ester
PAC	Polyacetylene
PAE	Polyarylether
PAEK*	Polyarylether ketone
PAI	Polyamidimide
PAMI	Polyaminobismaleinimide
PAN*	Polyacrylonitrile
PANI	Polyaniline, polyphenylene amine
PAR	Polyarylate
PARA	Polyarylamide
PARI	Polyarylimide
PB	Polybutene
PBA	Polybutylacrylate
PBI	Polybenzimidazole
PBMI	Polybismaleinimide
PBN	Polybutylene naphthalate
PBO	Polyoxadiabenzimidazole
PBT*	Polybutylene terephthalate
PC*	Polycarbonate (from bisphenol-A)
PCPO	Poly-3,3-bis-chloromethylpropylene oxide
PCTFE	Polychlorotrifluoro ethylene
PDAP	Polydiallylphthalate resin

Continued on next page

Acronym	Chemical notation
PDCPD	Polydicyclopentadiene
PE*	Polyethylene
PE-HD	Polyethylene – high density
PE-HMW	Polyethylene – high molecular weight
PE-LD	Polyethylene – low density
PE-LLD	Polyethylene – linear low density
PE-MD	Polyethylene – medium density
PE-UHMW	Polyethylene – ultra high molecular weight
PE-ULD	Polyethylene – ultra low density
PE-VLD	Polyethylene – very low density
PE-X	Polyethylene, crosslinked
PEA	Polyesteramide
PEDT	Polyethylenedioxythiophene
PEEEK	Polyetheretheretherketone
PEEK	Polyetheretherketone
PEEKEK	Polyetheretherketoneetherketone
PEEKK	Polyetheretherketoneketone
PEI*	Polyetherimide
PEK	Polyetherketone
PEKEEK	Polyetherketoneetheretherketone
PEKK	Polyetherketoneketone
PEN*	Polyethylenenaphthalate
PEOX	Polyethylene oxide
PESI	Polyesterimide
PES*	Polyethersulfone
PET*	Polyethylene terephthalate
PET-G*	Polyethylene terephthalate, glycol modied
PF*	Phenolic formaldehyde resin
PFMT	Polyperfluorotrimethyltriazine rubber
PFU	Polyfuran
PHA	Polyhydroxyalkanoate
PHB	Polyhydroxybutyrate
PHFP	Polyhexafluoropropylene
PI*	Polyimide

Continued on next page

Acronym	Chemical notation
PIB	Polyisobutylene
PISO	Polyimidsulfone
PK*	Polyketone
PLA	Polylactide
PMA	Polymethylacrylate
PMI	Polymethacrylimide
PMMA*	Polymethylmethacrylate
PMMI	Polymethacrylmethylimide
PMP	Poly-4-methylpentene-1
PMPI	Poly-m-phenylene-isophthalamide
PMS	Poly-a-methylstyrene
PNF	Fluoro-phosphazene rubber
PNR	Polynorbornene rubber
PO	Polypropylene oxide rubber
PO	General notation for polyolefins, polyolefin-derivates and -copolymers
POM*	Polyoxymethylene (polyacetal resin, polyformaldehyde)
PP*	Polypropylene
PPA	Polyphthalamide
PPB	Polyphenylenebutadiene
PPC	Polyphthalate carbonate
PPE*	Polyphenylene ether, old notation PPO
PPI	Polydiphenyloxide pyromellitimide
PPMS	Poly-para-methylstyrene
PPOX	Polypropylene oxide
PPP	Poly-para-phenylene
PPQ	Polyphenylchinoxaline
PPS*	Polyphenylene sulfide
PPSU*	Polyphenylene sulfone
PPTA	Poly-p-phenyleneterephthalamide
PPV	Polyphenylene vinylene
PPY	Polypyrrol
PPYR	Polyparapyridine
PPYV	Polyparapyridine vinylene
PS*	Polystyrene

Continued on next page

Acronym	Chemical notation
PSAC	Polysaccharide, starch
PSIOA	Polysilicooxoaluminate
PSS	Polystyrenesulfonate
PSU*	Polysulfone
PT	Polythiophene
PTFE*	Polytetrafluoroethylene
PTHF	Polytetrahydrofuran
PTT	Polytrimethyleneterephthalate
PUR*	Polyurethane
PVAC	Polyvinylacetate
PVAL	Polyvinylalcohol
PVB	Polyvinyl butyral
PVBE	Polyvinyl isobutylether
PVC*	Polyvinyl chloride
PVC/EVA	Polyvinyl chloride-ethylene vinylacetate
PVDC*	Polyvinylidene chloride
PVDF	Polyvinylidene fluoride
PVF	Polyvinyl fluoride
PVFM	Polyvinyl formal
PVK	Polyvinyl carbazole
PVME	Polyvinyl methylether
PVMQ	Polymethylsiloxane phenyl vinyl rubber
PVP	Polyvinyl pyrrolidone
PVZH	Polyvinyl cyclohexane
PZ	Phosphazene rubber with phenoxy groups
RF	Resorcin formaldehyde resin
SAN*	Styrene acrylonitrile
SB*	Styrene butadiene
SBMMA	Styrene butadiene methylmethacrylate
SBR	Styrene butadiene rubber
SBS	Styrene butadiene styrene
SCR	Styrene chloroprene rubber
SEBS	Styrene ethene butene styrene
SEPS	Styrene ethene propene styrene

Continued on next page

Acronym	Chemical notation
SEPDM	Styrene ethylene propylene diene rubber
SI	Silicone, Silicone resin
SIMA	Styrene isoprene maleic acid anhydride
SIR	Styrene isoprene rubber
SIS	Styrene isoprene styrene block copolymer
SMAB	Styrene maleic acid anhydride butadiene
SMAH*	Styrene maleic acid anhydride
SP	Aromatic (saturated) polyester
TCF	Thiocarbonyldiuoride copolymer rubber
TFEHFPVDF	Tetrafluoroethylene hexafluoropropylene vinylidene fluor
TFEP	Tetrafluoroethylene hexafluoropropylene
TM	Thioplastics
TOR	Polyoctenamer
TPA*	Thermoplastic elastomers based on polyamide
TPC*	Thermoplastic elastomers based on copolyester
TPE	Thermoplastic elastomers
TPE-A	see TPA
TPE-C	see TPC
TPE-O	see TPO
TPE-S	see TPS
TPE-U	see TPU
TPE-V	see TPV
TPO*	Thermoplastic elastomers based on olefins
TPS*	Thermoplastic elastomers based on styrene
TPU*	Thermoplastic elastomers based on polyurethane
TPV*	Thermoplastic elastomers based on crosslinked rubber
TPZ*	Other thermoplastic elastomers
UF	Urea formaldehyde resin
UP*	Unsaturated polyester resin
VCE	Vinylchloride ethylene
VCEMAK	Vinylchloride ethylene ethylmethacrylate
VCEVAC	Vinylchloride ethylene vinylacetate
VCMAAN	Vinylchloride maleic acid anhydride acrylonitrile
VCMAH	Vinylchloride maleic acid anhydride

Continued on next page

Appendix I

Acronym	Chemical notation
VCMAI	Vinylchloride maleinimide
VCMAK	Vinylchloride methacrylate
VCMMA	Vinylchloride methylmethacrylate
VCOAK	Vinylchloride octylacrylate
VCPAEAN	Vinylchloride acrylate rubber acrylonitrile
VCPE-C	Vinylchloride-chlorinated ethylene
VCVAC	Vinylchloride vinylacetate
VCVDC	Vinylchloride vinylidenechloride
VCVDCAN	Vinylchloride vinylidenechloride acrylonitrile
VDFHFP	Vinylidenechloride hexafluoropropylene
VF	Vulcanized fiber
VMQ	Polymethylsiloxane vinyl rubber
VU	Vinylesterurethane
XBR	Butadiene rubber, containing carboxylic groups
XCR	Chloroprene rubber, containing carboxylic groups
XF	Xylenol formaldehyde resin
XNBR	Acrylonitrile butadiene rubber, containing carboxylic groups
XSBR	Styrene butadiene rubber, containing carboxylic groups

Table I.2 Commonly Used Symbols Describing Polymer Characteristics

Symbol	Material characteristic
A	Amorphous
B	Block-copolymer
BO	Biaxially oriented
C	Chlorinated
CO	Copolymer
E	Expanded (foamed)
G	Grafted
H	Homopolymer
HC	Highly crystalline
HD	High density
HI	High impact
HMW	High molecular weight
I	Impact
LD	Low density
LLD	Linear low density
(M)	Metallocene catalyzed
MD	Medium density
O	Oriented
P	Plasticized
R	Randomly polymerized
U	Unplasticized
UHMW	Ultra high molecular weight
ULD	Ultra low density
VLD	Very low density
X	Cross-linked
XA	Peroxide cross-linked
XC	Electrically cross-linked

Appendix II Nomenclature

Roman Characters

A_2	Constant for determining the control limits
c	Number of occurrences of nonconformities in a sample
d_2	Constant from Table 7.2 in Chapter 7
D_3	Constant for determining control limits
D_4	Constant for determining control limits
D_r	Discrimination ratio
K	Number of intervals
L	Loss
n	Number of units in a sample
np	Number of nonconforming items in a sample of n items
N	Number of measurements
M	Mean target value
M_i	ith moment about the mean
p	Proportion of nonconforming items in a sample
r	Correlation coefficient
r^2	Coefficient of determination
r_i	Intraclass correlation coefficient
R	Range
$\bar{R}$	Average range
s	Sample standard deviation
u	Number of occurrences of nonconformities per production unit
v	Variance
$\bar{x}$	Sample mean
$\tilde{x}$	Median
$\hat{x}$	Mode
$\bar{\bar{x}}$	Grand average

Greek Characters

α	Alpha (alpha risk)
β	Beta (beta risk)
μ	Population mean
σ	Standard deviation
χ	Factor in chi-square test

Appendix III   SPC/DOE Software

III.1   Continuous Quality Improvement (CQI)

Traditional SPC is most commonly used for existing processes to determine if the process is in-control and to quantify the process capability. It does not tell us, however, whether the processing conditions are close to the optimum settings. In order to obtain this information, we have to go beyond traditional SPC. Several software packages aim to do this; some of these will be discussed next.

Streamer 1

One software package that goes beyond traditional SPC is Streamer1 from Synchrostat Corp. in Boulder, Colorado.

The first step in using Streamer1 is to construct a process model that describes the relationships between the process parameters. The tools that are used in this phase include regression analysis and design of experiments; these subjects are covered in Chapter 8 on advanced statistical tools. The process characterization establishes target values for all key process and product variables. Examples of key process variables are melt pressure, cavity pressure, and injection rate; key product variables are part weight, part appearance, and product dimensions.

The next step is to optimize the process. Streamer1 optimizes for product variables, process variables, or other factors, such as the financial loss associated with imperfect quality, also known as Taguchi loss. Most commonly, the optimization is focused on reducing product variation. Once the large variations have been eliminated, the process trends can be predicted. These are based on time-series predictions using an exponentially weighted moving average of process parameters. Thus, an early indication of a process drift can be obtained.

The package performs up to 300 tests on every variable for every parameter of every shot and provides results in real time, in other words results are displayed immediately. Internally, the P-values for each statistical result are determined to assess the statistical significance. This information is used to indicate whether the observed variation is random or may require corrective action. The probability levels associated with the statistical significance can be set by the user.

The underlying math of the Streamer1 software makes no assumptions regarding the distribution of the data. The software can provide not only the traditional Gaussian CpK, but also several non-Gaussian (non-normal) CpK measures. The software operates in an asynchronous mode; thus, processing events can trigger responses when they occur, rather than when the computer polls the particular parameter(s).

Pro-T-coN

This software package was developed by GS Technology in England and was licensed to Syscon Planstar in the US. Pro-T-coN performs a process analysis based on process data entered into the computer. The process variables have to be categorized as controllers, derivatives, or resultants. A controller is a variable that is directly adjustable, such as screw speed. A derivative is a dynamic variable that has a value only when the process is running, for instance cavity pressure. Derivatives are not directly controllable, but are a consequence of controllers. Resultants are process outputs, such as weights, dimensions, and surface conditions.

The program will accommodate up to 100 variables. Targets have to be set for the resultants; there can be up to 15 targets. After the process data is entered into the computer, the process model is determined. The program calculates the best target values that can be achieved and what control settings are necessary to produce these targets. Also available is information on the percentage of total variability captured for each target. Values above 90 percent indicate that most significant factors affecting the target have been included; values below 90 percent indicate a substantial probability that a significant factor has been omitted.

A Pareto plot shows the contribution made by each variable to the total variation in a particular target. Another screen shows the ratio listing the effect of process variables on target values. Successive calculations will yield information on the best target tolerance that can be obtained in real life. The end result is the SMI, the standard manufacturing instructions; this is a set of process conditions that will produce an acceptable product.

G2, GDA, and GSPC

G2 by Gensym Corp., Cambridge, Mass., combines object-oriented modeling and animation to create graphical models that represent the dynamic process behavior. G2 Diagnostic Assistant, GDA, is a graphical tool that allows process engineers to build intelligent process management applications. With G2 and GDA combined with GSPC add-ons, users can access a library of graphical software components to configure online SPC solutions quickly. SPC charts developed with GSPC can be linked directly to process management applications for real-time quality management in discrete, batch, and continuous manufacturing operations. SPC control charts can be represented using graphical control chart blocks, and then can be connected to graphical representations of a manufacturing operation. Blocks can pass data values, logic states, and other objects along connection paths.

Potential process problems can then be detected by accumulating deviations between process measurements and expected values. The deviations can be plotted on control charts, which can display and analyze patterns in real time using special pattern observation blocks together with logic networks. Thus, real-time SPC can be linked with root-cause analysis to pinpoint the source of a quality problem. G2 Version 8.3 is available for Windows, Solaris SPARC, HP-UX, HP Tru64, IBM AIX and Linux Red Hat Operating Systems. It is Web enabled and includes support for web services, sending and receiving messages based on SOAP and HTTP.

SPM, Statistical Predictive Maintenance

This package by Infor Alpharetta, GA, is used to organize maintenance schedules. The software can establish trends for equipment use by tracking and analyzing variables like vibrations, temperature, current, pressure, etc. By comparing these trends to specified tolerances and limits, the software can automatically schedule maintenance or alert an operator to a possible breakdown.

Autoset

This software package by Bykom, St. Peters, Mo., uses neural network technology to learn a process and then, with the push of a button, directly links process setpoints to quality assurance laboratory results. When this relationship is determined, the program predicts QA lab results with its "virtual inspector." For any process Autoset can predict up to four product quality attributes such as moisture, thickness, density, surface quality, etc. The program also contains a setpoint generator that automatically adjusts setpoints online in real time to optimize the process, eliminating variations before they can create problems. Autoset allows even the production operator to build solutions to complicated problems. Packages like Autoset will make neural network solutions easier to obtain in a wide range of manufacturing operations.

As mentioned earlier, several software packages are commercially available incorporating advanced statistical techniques such as design of experiments, regression analysis, artificial intelligence, expert systems, fuzzy logic, neural networks, etc. It goes beyond the purpose of this text to discuss these in detail. However, it can be expected that these programs will proliferate in the near future and that some of these will be integrated with new injection molding machines, allowing easier optimization of the process and computer based problem solving by expert systems based on artificial intelligence. In fact, several manufacturers of injection molding machines are actively working in this area.

III.2 Software Packages for Statistical Analysis

Statistica for PC and Mac, does multiple regression, general ANCOVA/MANCOVA, quality control, process analysis, experimental design, and more.
By Statsoft, 2325 E. 13th Street, Tulsa, OK 74104, www.statsoft.com

Statgraphics for PC, does ANOVA, regression analysis, quality control, experimental design, and more.
By STSC, Inc., 2115 East Jefferson Street, Rockville, MD 20852, www.statgraphics.com

NWA Quality Analyst for PC does standard statistical control charting, both variable and attribute control charts, and process capability.
By Northwest Analytical, 520 N.W. Davis Street, Portland, OR 97209, www.nwasoft.com

S-Plus for Unix and DOS systems with at least 386 processor and 387 math coprocessor, has over 1200 functions for performing exploratory data and statistical analysis, includes ANOVA, linear and nonlinear models, stepwise regression, etc.
By Statistical Sciences, 1700 Westlake Avenue, Suite 500, Seattle, WA 98109

Systat for PC and Mac has statistics, graphics, and data management capability. Statistics include linear, polynomial, stepwise, weighted regression with extended diagnostics, ANOVA, ANCOVA, MANOVA, etc.
By Systat, Inc., 1800 Sherman Avenue, Evanston, IL 60201-3793, www.systat.com

Design-Ease helps experimenters improve their product or process. It includes two-level factorial, fractional factorial, and Plackett-Burman designs. Also offered Design-Expert, which draws response surface maps that lead directly to optimum performance.
By Stat-Ease, Inc., 2021 East Hennepin Avenue, Minneapolis, MN 55413-2723, www.statease.com

Minitab Statistical Software for PC and Mac is used for organizing, analyzing, and reporting statistical data; it does SPC, simple and multiple regression analysis, design of experiment, and more.
By Minitab, Inc., 3081 Enterprise Drive, State College, PA 16801, www.minitab.com

JMP Statistical Discovery Software for PC and Mac does multivariate tests, Shewhart control charts, non-linear fitting, and more. Packages for other computer systems also available.
By SAS Institute, SAS Campus Drive, Cary, NC 27330, www.jmp.com

BMDP *New System* for Windows PCs does stepwise regression, spectral analysis, Box-Jenkins time series, multivariate ANOVA and ANCOVA, and more.
By Statistical Solutions, Stonehill Corporate Center, 999 Broadway, Saugus, MA 01906, www.statsol.ie

Applied Stats for the PC and Mac. Software is used for data acquisition, both variables and attributes data, SPC functions: variables, attributes, and short runs, scatter plots, capability indices, histograms, Pareto charts, gauge capability studies, etc.
By ASI DataMyte, Royal Oak, Michigan, 222 East Fourth St., www.asidatamyte.com

MPC (Multistage Process Control) quality-assurance program designed for a wide range of single and multi-stage operations in plastics processing. The software can be used for the systematic analysis, optimization and monitoring of the quality of different plastics processing operations and molded parts at the same time. CPC is used to evaluate all the parts produced without having to inspect each part individually.
By KTP (Institute of Polymer Engineering), University of Paderborn, Warburger Strasse 100, D-33098 Paderborn, Germany, www.ktpweb.de

Quantum is used for data acquisition and analyzing both variables and attributes data. Provides real-time shop floor data collection, charting, data analysis and alarm notification to the floor operator, assignable cause prompts and automatic trend analysis.
By ASI DataMyte, Royal Oak, Michigan, 222 East Fourth St., www.asidatamyte.com

ONQUALITY is a real-time quality management system that integrates SPC quality data from receiving, processing, manufacturing, and the lab to provide complete SPC and tracking of the production process. Based on user-definable SQL database, data from any source can be used to present a true quality view of the process.
By Automation ONSPEC Software, Inc., P.O. Box 743, Rancho Cordova, CA 95741-0743, http://automationonspec.com

SPC/PI+ can perform a complete SPC analysis with basic and advanced control charts (CUSUM and Exponentially Weighted Moving Average chart), full chart tool bar for magnify, scroll, data ID, exclude and adding comments to each data point, auto pattern analysis tool button, deals with autocorrelation, gage capability, and real time SPC through RS232 port.
By Qualitran Professional Services, PO Box 295 Stroud, Ontario, L0L2M0, Canada, www.qualitran.com

SPC-Light creates studies that include variables or attributes or combinations of the two, charts show multiple variables side by side, this shows the process spread relative to the variable tolerance, presents the data alongside the chart, view an image of the product or link the SPC study to an external document, produces Pareto of Causes & Actions.
By Lighthouse Systems Inc., Building 3, 6780 Pittsford-Palmyra Road, Fairport NY 14450, www.lighthousesystems.com

eSPC web-based, fully hosted SPC solution that collects and analyzes quality data in real-time, test data input into the system during production, software presents the results on customized charts and graphs, and automatically alarms users when processes go out of control, communicates with the server using standard HTTP protocol and 128 Bit encryption techniques.
By InfinityQS International Inc., 14900 Conference Center Drive Chantilly, VA 20151, www.infinityqs.com

Visiprise Real-Time SPC gathers product and process-related information in real-time, proactive statistical analysis in real-time — send alarms/alerts based on configurations and limits to the proper personnel, provide immediate feedback — during the manufacturing

process and data collection, support and display SPC charts — quickly identify alarming trends to make informed decisions.
By Visprise, Alpharetta, 2500 Northwinds Parkway, Suite 500, GA 30004,
www.visiprise.com

WinSPC software provides real-time, intelligent process feedback to monitor in-line production and quality data, performs advanced statistical analysis, creates and distributes quality reports for corporate, customers, and suppliers, automated closed-loop process feedback, provides optional built-in audit tracking for original configuration and specifications.
By DataNet Quality Systems, 24567 Northwestern Hwy., Southfield, MI 48075,
www.winspc.com

DataLyzer® Spectrum a comprehensive, real-time, Windows® SPC software package simplifying the SPC tasks data collection and charting. Analysis is comprehensive intended to communicate the truth about the process, user package can stand-alone or form the foundation of a larger local area network scenario.
By Stephen Computer Services, Inc., 1857 E. West Maple Rd. Walled Lake, MI 48390,
www.datalyzer.com

Table III.1 Data Acquisition/Statistical Software Packages.

Supplier	DA	DOE	GR&R	SPC
Air Gage Co.	●		●	●
Alligator Technologies	●			●
Am. Inst. Quality & Reliability	●		●	●
Am. Quality and Prod. Software Co.	●		●	●
American Systems, Inc.	●		●	●
Applied Statistics, Inc.	●		●	●
Automated Technology Associates	●	●	●	●
ASI Data Myte	●			●
BBN Software Products	●	●	●	●
Brunswick Instrument	●			●
Collins Data Systems, Inc.	●		●	●
Compusense	●	●		●
Concert Corp.	●			●
Cooperative Systems Co.	●		●	●
Cybermetrics Corp.	●		●	●
DataNet Technologies	●		●	●
DES Systems	●			●
Dimensional Data Systems	●			●
Fred. V. Fowler Co, Inc.	●		●	●
GageTalker Corp.	●		●	●
Golden Coast Software, Inc.	●			●
The Harrington Group, Inc.	●			●
Hertzler Systems, Inc.	●		●	●
InfinityQS	●			●
Intaq, Inc.	●			●
Intercim Corp.	●		●	●
International Quality Institute	●			●
International Qual-Tech, Ltd.	●	●	●	●
IQS, Inc.	●	●	●	●
Jonathon Cole Associates, Inc.	●			●
John A. Keane and Associates, Inc.	●	●		●
Kurt Manufacturing Co.	●		●	●
Lionheart Press, Inc.	●	●		●
Major Micro Systems, Inc.	●			●

DA = data acquisition, DOE = design of experiment, GR&R = gage reproducibility and repeatability, SPC = statistical process control

Table III.2 Suppliers of DA/SPC/DOE Software

Supplier	DA	DOE	GR&R	SPC
MetriStat Div. Business Syst. Design	•			•
Mitutoyo – SPC Group	•		•	•
Murphy Software Co.	•		•	•
Northwest Analytical, Inc.	•			•
Penfact, Inc.	•			•
Perry Johnson, Inc.	•	•		•
Pilgrim Software, Inc.	•		•	•
Pister Group, Inc.	•	•	•	•
PQ Systems, Inc.	•	•	•	•
Process Integrity, Inc.	•			•
Qualitran Professional Services, Inc.	•			•
Quality Systems, Inc.	•	•	•	•
Quality America, Inc.	•	•		•
Quality Measurement Systems	•			•
Quality Resources	•	•	•	•
Salerno Manufacturing Systems, Inc.	•		•	•
SAS Institute, Inc.	•	•		•
Solution Specialists	•	•	•	•
StaScan, Inc.	•			•
Stephen Computer Services, Inc.	•			•
Stochos, Inc.	•	•	•	•
Synchro Stat Systems Corp.	•	•	•	•
Teque, Inc.	•	•		•
The Crosby Co.	•		•	•
Training Technologies, Inc.	•			•
Zontec, Inc.	•			•

DA = data acquisition, DOE = design of experiment, GR&R = gage reproducibility and repeatability, SPC = statistical process control

Appendix IV Expressions for the Normal Distribution Curve and z-Table

The equation for the standard normal distribution curve is (see Fig. IV.1):

$$P(x) = \frac{\exp[-(x-\mu)^2/(2\sigma)^2]}{\sigma\sqrt{2\pi}} \tag{IV.1}$$

where μ is the mean and σ the standard deviation of the normal distribution.

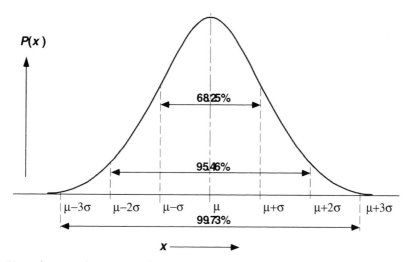

Figure IV.1 The normal or Gaussian distribution

The chance that a measurement is larger than $x = \mu + z\sigma$ is given by:

$$F(x \geqslant \mu + z\sigma) = \int_{\mu+z\sigma}^{\infty} P(x)dx \tag{IV.2}$$

This is the area under the $P(x)$ curve from $x = \mu + z\sigma$ to $x = \infty$ (infinity). Since $P(x)$ is symmetric about $x = \mu$, $F(x \geq \mu + z\sigma)$ is equal to $F(x \leq \mu - z\sigma)$ or:

$$\int_{\mu+z\sigma}^{\infty} P(x)dx = \int_{-\infty}^{\mu-z\sigma} P(x)dx \tag{IV.3}$$

The total area under the curve is 1; thus:

$$\int_{-\infty}^{\infty} P(x)dx = 1 \tag{IV.4}$$

Data on $F(x \geq \mu + z\sigma)$ or $F(x = \mu \text{ to } x = \mu + z\sigma)$ are often presented in tabular form; an example is the z table (Table IV.1) where $z = (x - \mu)/\sigma$. The table shows the area under the normal curve from $z = 0$ to z or $F(z = 0 \text{ to } z)$, which the same as $F(x = \mu \text{ to } x = \mu + z\sigma)$; see also Fig. IV.2.

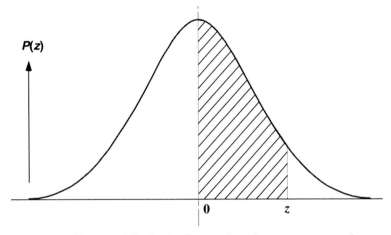

Figure IV.2 Fraction of the normal distribution from $z = 0$ to z (or $x = \mu$ to $x = \mu + z\sigma$)

Table IV.1 z Table, Area under the Normal Distribution Curve from 0 to z

Normal Deviate z	.00	.01	.02	.03	.04	.05	.06	.07	.08	.09
0.0	.0000	.0040	.0080	.0120	.0160	.0199	.0239	.0279	.0319	.0359
0.1	.0398	.0438	.0478	.0517	.0557	.0596	.0636	.0675	.0714	.0753
0.2	.0793	.0832	.0871	.0910	.0948	.0987	.1026	.1064	.1103	.1141
0.3	.1179	.1217	.1255	.1293	.1331	.1368	.1406	.1443	.1480	.1517
0.4	.1554	.1591	.1628	.1664	.1700	.1736	.1772	.1808	.1844	.1879
0.5	.1915	.1950	.1985	.2019	.2054	.2088	.2123	.2157	.2190	.2224
0.6	.2257	.2291	.2324	.2357	.2389	.2422	.2454	.2486	.2518	.2549
0.7	.2580	.2612	.2642	.2673	.2704	.2734	.2764	.2794	.2823	.2852
0.8	.2881	.2910	.2939	.2967	.2995	.3023	.3051	.3078	.3106	.3133
0.9	.3159	.3186	.3212	.3238	.3264	.3289	.3315	.3340	.3365	.3389
1.0	.3413	.3438	.3461	.3485	.3508	.3531	.3554	.3577	.3599	.3621
1.1	.3643	.3665	.3686	.3708	.3729	.3749	.3770	.3790	.3810	.3830
1.2	.3849	.3869	.3888	.3907	.3925	.3944	.3962	.3980	.3997	.4015
1.3	.4032	.4049	.4066	.4082	.4099	.4115	.4131	.4147	.4162	.4177
1.4	.4192	.4207	.4222	.4236	.4251	.4265	.4279	.4292	.4306	.4319
1.5	.4332	.4345	.4357	.4370	.4382	.4394	.4406	.4418	.4429	.4441
1.6	.4452	.4463	.4474	.4484	.4495	.4505	.4515	.4525	.4535	.4545
1.7	.4554	.4564	.4573	.4582	.4591	.4599	.4608	.4616	.4625	.4633
1.8	.4641	.4649	.4656	.4664	.4671	.4678	.4686	.4693	.4699	.4706
1.9	.4713	.4719	.4726	.4732	.4738	.4744	.4750	.4756	.4761	.4767
2.0	.4772	.4778	.4783	.4788	.4793	.4798	.4803	.4808	.4812	.4817
2.1	.4821	.4826	.4830	.4834	.4838	.4842	.4846	.4850	.4854	.4857
2.2	.4861	.4864	.4868	.4871	.4875	.4878	.4881	.4884	.4887	.4890
2.3	.4893	.4896	.4898	.4901	.4904	.4906	.4909	.4911	.4913	.4916
2.4	.4918	.4920	.4922	.4925	.4927	.4929	.4931	.4932	.4934	.4936
2.5	.4938	.4940	.4941	.4943	.4945	.4946	.4948	.4949	.4951	.4952
2.6	.4953	.4955	.4956	.4957	.4959	.4960	.4961	.4962	.4963	.4964
2.7	.4965	.4966	.4967	.4968	.4969	.4970	.4971	.4972	.4973	.4974
2.8	.4974	.4975	.4976	.4977	.4977	.4978	.4979	.4979	.4980	.4981
2.9	.4981	.4982	.4982	.4983	.4984	.4984	.4985	.4985	.4986	.4986
3.0	.49865	.4987	.4987	.4988	.4988	.4989	.4989	.4989	.4990	.4990
4.0	.49997									

Appendix V   Glossary

Accuracy. The closeness of agreement between an observed value and an accepted reference value.

Advanced Statistical Methods. More sophisticated techniques of statistical process analysis and control than included in basic statistical methods; this can include more advanced control chart techniques, regression analysis, design of experiments, advanced problem solving techniques, etc.

Alpha risk. The risk of rejecting the null hypothesis when the hypothesis is actually correct.

Assignable cause. A significant source of variation that is not inherent to the process but has an identifiably reason. Also called special causes, sporadic causes, chaotic causes, and unnatural causes.

Attribute. A product characteristic than can only be described by one of two possible conditions.

Auto-Tuning Controllers. Controllers that can automatically determine and set the appropriate values of the tuning parameters of the controller.

Average. See *Mean*.

Basic Statistical Methods. Applies the theory of variation through use of basic problem-solving techniques and statistical process control; includes control chart construction and interpretation for both variables and attributes data and capability analysis.

Beta risk. The risk of accepting the null hypothesis when, in fact, the hypothesis is incorrect.

Bias. The difference between the true value and the average of repetitive measurements made by a measurement process.

Binomial Distribution. A discrete probability distribution for attributes data that applies to conforming and nonconforming units and underlies the p and np charts.

Capable Process. A process that can deliver in spec. products consistently with a high statistical significance, see also process capability.

Capability Index. The ratio of the tolerance to the process variation; usually the process variation is expressed as 6σ.

Cause and Effect Diagram. A simple tool for individual or group problem-solving that uses graphic description of the various process elements to analyze potential sources of process variation. Also called fishbone diagram (after its shape) or Ishikawa diagram (after its developer).

Cause Listing Diagram. A diagram showing the various possible causes of a problem.

Cavity. The hollow space in the mold with the shape of the injection-molded part.

c-**Chart.** A chart showing the number of defects or nonconformities. It requires constant subgroup size.

Central Limit Theorem. Sample averages (or x-bars) will follow a normal distribution as long as only common cause variations occur.

Central Line. The line on a control chart that represents the average value of the item being plotted.

Central Tendency. The tendency of data to group around a certain value. Measures of central tendency are mean, mode, and median.

Chance Cause. See *Common Cause*.

Chaotic Cause. See *Assignable Cause*.

Characteristic. A distinguishing feature of a process or its output on which variables or attributes data can be collected.

Chi-square (χ^2) test. Test of normality.

Chronic cause. See *Common Cause*.

Cluster. A region in the control chart where points are closely grouped together.

Common Cause. A source of variation that is inherent to the process. Also called random causes, chronic causes, and natural causes.

Consecutive. Units of output produced in succession; a basis for selecting subgroup samples.

Control Chart. A graphical method for determining whether a process is in a state of statistical control.

Control Chart Method. The method of using control charts to determine whether or not processes are in a stable state.

Control Limit. A line on a control chart used as a basis for judging the stability of the process; not to be confused with specification limit.

Correlation Table. A table showing the relationship between two variables.

CP. See *Process Potential Index*.

CpK. See *Capability Index*.

CR. The CR index is the inverse of CP.

CuSum. An advanced statistical method that uses current and recent past process data to detect small to moderate shifts in the process average or variability. CuSum stands for cumulative sum of deviations from the target and puts equal weight on the current and recent past data.

Cycle. A regular, repeating pattern on a control chart characterized by alternating regions of increasing and decreasing values of the data plotted.

Defect. A departure of a quality characteristic from its intended level or state.

Degradation. Molecular deterioration of materials such as resins, organic fibers or polymers because of overheating.

Defective Units. A unit of product or service containing at least one defect, or having several imperfections that combined cause the unit not to satisfy intended normal, or reasonably foreseeable usage requirements.

Design of experiments. DOE is a structured, organized method for determining the relationship between factors affecting a process and the output of that process.

Deviation. The difference between one of a set of numbers and its stated value or intended level.

Devolatilization. The removal of volatile components.

Differential scanning calorimetry. A technique in which the difference in the amount of heat required to increase the temperature of a sample and reference are measured as a function of temperature.

Dispersive mixing. A mixing process in which an intrinsic change takes place in the physical character of one of the components. Agglomerates are reduced in size by fracture due to stresses generated during mixing.

Distribution. A method of describing the variation of a stable system, in which individual values are not predictable but in which the outcome as a group form a pattern that can be described in terms of its location, spread, and shape.

Distributive mixing. Reducing the composition non-uniformity where the ingredients do not exhibit a yield stress.

Event. The occurrence of some attribute.

Fishbone Diagram. See *Cause and Effect Diagram*.

Freaks. Points located outside the control limits of a control chart.

Gage R&R. The gage repeatability and reproducibility.

Gaussian Distribution. See *Normal Distribution*.

GOC. See *Gravimetric Output Control*.

Gravimetric Output Control. A method of controlling the extruder output by measuring the weight of the material discharged from the feed hopper to the extruder.

Histogram. A bar chart with the height of each bar indicating how many data occur within a certain interval; the width of the bar equals the width of the interval.

Hysteresis. The failure of a property that has been changed to return to its original value when the cause of the change is removed.

Individual. A single unit, or a single measurement of a characteristic, often denoted by the symbol x.

Inspection. The process of measuring, examining, testing, or otherwise comparing the unit with the applicable requirements.

Ishikawa Diagram. See *Cause and Effect Diagram*.

Item. An object or quantity of material on which a set of observations can be made.

Kurtosis. A characteristic of the shape of a distribution curve where there is a relatively high concentration in the middle and out on the tails with a relatively steep drop in between.

Lower Specification Limit or LSL. The lowest value of a characteristic that the specifications will allow.

LSL. See *Lower Specification Limit*.

Mean. The sum of the observations divided by the number of observations. Also called average.

Median. The middle measurement of an ordered set of numbers with the numbers arranged in order of size. For an even set of numbers, it is the average of the two middle numbers.

Metrology. The science of measurement.

Mixture. A condition of a control chart where there is a relative absence of points near the central line.

Mode. The most frequently occurring value in a set of data.

Moving Range. The difference in value between two or more, most recent, successive samples.

Multimodal Distribution. A distribution with more than one peak.

Natural Cause. See *Common Cause*.

Negative Skew. A distribution whose peak is on the upper side of the distribution curve.

Nonconforming Units. Units which do not conform to a specification or other inspection standard; sometimes called discrepant or defective units. p and np charts are used to analyze systems producing nonconforming units.

Nonconformity. A specific occurrence of a condition which does not conform to a specification or other inspection standard; sometimes called discrepancy or defect.

Non-linearity. The largest deviation of the actual measurements from the straight line characteristic divided by the full scale.

Normal distribution. A bell-shaped distribution described by the formula in Appendix IV. Also called Gaussian distribution.

Null hypothesis. An initial hypothesis about the population that is tested by drawing and analyzing samples from the population.

np-Chart. A chart for the actual number of defective items in the subgroup. It requires a constant subgroup size.

On/Off Temperature Control. Temperature control where heating or cooling is applied either full on or completely off with no intermediate values.

Pareto Chart. A simple tool for problem solving that involves ranking all potential problem areas or sources of variation according to their contribution to cost or total variation.

***p*-Chart.** An attributes control chart for the percentage of defective items in a subgroup when the subgroup size is not constant.

PID Control. See *Proportional/Integral/Derivative Control*.

Plasticating Extruder. An extruder that receives solid material and melts or plasticates the material as it is conveyed to the discharge.

Poisson Distribution. A discrete probability distribution for attributes data that applies to nonconformities and underlies the c and u control charts.

Population. The totality of items or units of material under consideration. Also called universe.

Positive Skew. A distribution whose peak is on the lower side of the distribution curve.

Precision. The extent to which repetitive measurements on a single unit agree.

Prevention. A future-oriented strategy that improves quality and productivity by directing analysis and action toward correcting the process itself.

Probability Plot. A plot of the cumulated distribution on probability paper.

Problem-Solving. The process of determining the causes of problems, as evidenced by certain symptoms, and taking action to eliminate the problem or reduce its severity.

Process. The combination of people, equipment, materials, methods, and environment that produce output — a certain product or service.

Process Analysis Diagram. A diagram showing all the steps in the process.

Process Capability. The ratio of the tolerance (USL-LSL) to the process variation (6σ).

Process Potential Index. Symbolized by CP, is the ratio of the tolerance to 6 times the process standard deviation. It is a measure of spread of the process data.

Proportional/Integral/Derivative Control. Control that uses proportional action combined with integral (reset) and derivative (rate) action.

PVT Diagram. Diagram that shows the equation of state that connects pressure, specific volume and temperature for a fluid.

Quality. The totality of features and characteristics of a product or service that bear on its ability to satisfy a given need.

Random cause. See *Common Cause*.

Randomness. A condition in which individual values are not predictable, although they may come from a definable distribution.

Random Sampling. The process of selecting units for a sample of size n in such a manner that all combinations of n units under consideration have an equal chance of being selected as the sample.

Range. The difference between the largest and smallest measurement.

Rational Subgroup. A subgroup gathered in a manner as to give the maximum chance for the measurements in each subgroup to be alike and the maximum chance for the subgroups to differ one from the other.

Repeatability. The variation obtained when one person measures the same quantity several times using the same measuring instrument.

Reproducibility. The variation in measuring averages due to differences among instruments, operators, etc.

Resolution of a Measuring Instrument. The smallest unit of measure which an instrument is capable of indicating.

Run. A consecutive number of points consistently increasing or decreasing or above or below the central line. This can be evidence of the presence of special causes of variation.

Sample. One of several individual pieces or measurements that is collected for analysis. In process control applications, sample is a synonym for subgroup.

Sample Size. The number of units in a sample.

Sampling Interval. In systematic sampling, the fixed interval of time, output, running hours, etc. between samples.

Scatter Diagram. A simple graph showing the cause on the horizontal axis and the effect on the vertical scale, used to assess the scatter in the data.

SEC. See *Specific Energy Consumption*.

Sensitivity of a Measuring Instrument. The smallest change in the measured quantity which the instrument is capable of detecting.

Shewhart control chart. See $\bar{x}$ *and R chart*.

Shewhart. Dr. Walter A. Shewhart developed basic methods for statistical monitoring of manufacturing processes.

Sigma (σ). The Greek letter used to designate the deviation of a population.

Signal to noise ratio. Ratio of a signal power to the noise power corrupting the signal.

Skew. Asymmetrical.

Skewness. The degree to which a distribution is asymmetrical.

SPC. See *Statistical Process Control*.

Special cause. See *Assignable Causes*.

Specification. The engineering requirement for judging acceptability of a particular characteristic.

Specific Energy Consumption. The amount of energy used per unit volume or unit mass of material.

Sporadic Causes. See *Assignable Causes*.

Spread. The span of values from the smallest to the largest in a distribution.

Stability. In process control: the absence of special causes of variation; the state of being in statistical control.

Stability. In measurement analysis: the difference in the average of at least two sets of measurements obtained with the same measuring device on a single unit taken at different times.

Standard Deviation. A measure of variability (dispersion) of observations that is the positive square root of the population variance.

Starve Feeding. A method of feeding an extruder where the feed material is metered in at a rate below the full intake capability of the extruder. In this case, the extruder output is determined by the feeding device, not by the screw speed of the extruder.

Statistic. A quantity calculated from a sample of observations, most often to form an estimate of some population parameter.

Statistical Process Control. The use of statistical methods to monitor and control a process.

Statistics. Deals with the collection, organization, analysis, and interpretation of numerical data.

Stratification. A condition of a control chart where 15 or more points are very close to the central line.

Subgroup. One or more events or measurements used to analyze the performance of a process.

Tolerance. The upper limit specification, USL, minus the lower specification limit, LSL.

Tool Control Chart. A control chart that allows current process data to be accurately compared to historical data or to correct data for possible bias.

***u*-Chart.** A chart showing the number of defects or nonconformities. It is used when the size of the subgroup varies.

Unimodal Distribution. A distribution with a single peak.

Universe. See *Population*.

Unnatural Cause. See *Assignable Cause*.

Upper Specification Limit. The maximum value of a characteristic that the specifications will allow.

USL. Acronym for upper specification limit.

Variable. A product characteristic that can be one of many possible values.

Variables Data. Quantitative data, where measurements are used for analysis.

Variance. The mean of the squared deviations from the arithmetic mean.

Variation. The inevitable differences among individual outputs of a process; the sources of variation can be grouped into two major classes: common causes and special causes.

Variation Analysis Diagram. A diagram that list the various possible sources of variation.

Weibull distribution. If the life of an item has the density function

$$f(X) = \frac{\beta(X-1)^{\beta-1} \exp[-(X-\gamma)^{\beta/\alpha}]}{\alpha} \quad \text{for } X \geq \gamma$$

$f(X) = 0$ elsewhere
with $\alpha > 0$ and $\beta > 0$, then X is said to have a Weibull distribution.

Weld Line. A weak line that can form in a plastic part as a result of insufficient entanglements of the molecules. This can happen when the plastic splits and recombines around a spider support in an extrusion die.

Wilks method. A method of determining process capability for a non-normal distribution.

$\bar{x}$ and R chart. Chart showing the average, $\bar{x}$, and range, R, of groups of data. Also called Shewhart control chart.

x and Rm chart. Chart showing individual measurements, x, and the moving range, Rm.

$\bar{x}$ and s chart. Chart showing the average, $\bar{x}$, and the standard deviation, s.

Zone Analysis. A method of detailed analysis of a Shewhart control chart which divides the x-bar chart between the control limits into three equidistant zones above and below the mean. These zones are sometimes referred to as "sigma" zones.

Appendix VI   Monitoring Systems for Injection Molding

The following is a tabular listing based on [33].

Company	Single IM	Multiple IMs	Process	Scheduling	MRP	Inventory	SPC	Other
Absolute Haitian	●							
Arburg, Inc.	●	●	ALS-AQC				●	AQS
Autojectors, Inc.	●	●	Turbo touch				●	
Barco Inc./Automation	●	●	●	●	●	●	●	PLANT-MASTER (MES)
Boy Machines Inc.	●	●		●			●	Transfer of data records
CMC Technologies	●	●	F, P, T, x, t				●	
DME Co.	●		P, T, stroke				●	
Engel	●	●	●	●			●	Micrograph Microtemp
Gefran Inc.	●	●	P, T, v, x					
Helicoid Instruments	●		●					
Hettinga Equipment		●					●	
HPM Corporation		●	●	●			●	
Hunkar Laboratories	●	●	●	●			●	
Husky IM Systems Ltd.	●	●	All CIM systems					
IMS Company	●	●	●					
JSW PMI	●							
Kraus Maffei Corp.	●	●		●	●	●	●	
Mattec Corp.	●	●	●	●			●	Material-use forecast

Continued on next page

Continued

Company	Single IM	Multiple IMs	Process	Scheduling	MRP	Inventory	SPC	Other
Meiki America, Inc.	•		•	•		•	•	•	Injection speed, P
MIR USA Corp.	•	•		•				•	
New Pacific Machinery	•		Digital computer						
Nicollet Process Engineering	•		PCA					•	Diagnostic functions, 4
Nissei America, Inc.	•	•	NC-NET			•		•	Transfer of data records
RJG Technologies	•	•	*	•				•	
Sumitomo			iiiQ-System						
Syscon-Plantstar	•	•	•	•		•	•	•	
TMC Instruments Inc.	•	•	Temp. Pressure						
Toshiba Machine Co. America	•	•	•						
Ube Machinery, Inc.	•	•						•	
Welltec USA, Inc.	•	•						•	

IM = injection molding machine
F = force
P = pressure
T = temperature
x = position
t = time
v = speed

* Cavity pressure, hydraulic pressure, fill time, cycle time, stroke, mold coolant flow rate, area-under-the-curve analysis, wave form interpretation.

Appendix VII   Conversion Constants

Length

1 kilometer	km	= 1E3 m
1 centimeter	cm	= 1E-2 m
1 millimeter	mm	= 1E-3 m
1 micron	mm	= 1E-6 m
1 inch	in	= 2.54E-2 m
1 milliinch	mil	= 2.54E-5 m
1 foot	ft	= 0.3048 m
1 mile	ml	= 1609 m

Mass

1 ton (metric)	t	= 1 E3 kg
1 gram	gr	= 1E-3 kg
1 ounce	oz	= 2.83E-2 kg
1 pound	lb	= 0.4536 kg
1 ton (US)	tn	= 907 kg
1 ton (UK)	ton	= 1016 kg

Force

1 dyne	dyn	= 1E-5 N
1 kg-force	kgf	= 9.81 N
1 ton-force	tf	= 9810 N
1 pound-force	lbf	= 4.448 N

Viscosity

| 1 poise | P | = 0.1 Pas |

Volume

1 cubic decimeter	dm^3	= 1E-3 m^3
1 cubic centimeter	cm^3	= 1E-6 m^3
1 liter	l	= 1E-3 m^3
1 cubic inch	in^3	= 1.639E-5 m^3
1 cubic foot	ft^3	= 2.832E-2 m^3
1 barrel	brl	= 0.159 m^3
1 gallon (US)	gal US	= 3.785E-3 m^3
1 gallon (UK)	gal UK	= 4.546E-3 m^3

Density

1 gram/CC	gr/cc	= 1E-3 kg/m^3
1 pound/ft^3	lb/ft^3	= 16.02 kg/m^3
1 pound/in^3	lb/in^3	= 0.0277 kg/m^3
1 gram/cc	gr/cc	= 0.0361 lb/in^3

Stress

1 megapascal	MPa	= 1E6 Pa
1 dyne per cm^2	dyn/cm^2	= 0.1 Pa
1 Newton per m^2	N/m^2	= 1 Pa
1 atmosphere	atm	= 1.013E5 Pa
1 bar	bar	= 1E5 Pa
1 pound per in^2	psi	= 6890 Pa
1 megapascal	MPa	= 145 Pa

Power

1 kilowatt	kW	= 1000 W
1 horsepower	hp	= 746 W
1 footpound/s	ftlbf/s	= 1.356 W

Specific Energy

1 calorie per gram	cal/gr	= 4190 J/kg
1 Btu per pound	Btu/lb	= 2326 J/kg
1 hphr per pound	hphr/lb	= 5.92E6 J/kg
1 kWhr per kilogram	kWhr/kg	= 3.60E6 J/kg
1 kWhr per kilogram	kWhr/kg	= 0.608 hphr/lb

1 poise	P	= 1 dyns/cm^2
1 lbfs/in^2	psis	= 6897 Pas
1 poise	P	= 1.45E-5 psis

Energy

1 newtonmeter	Nm	= 1 J
1 wattsecond	Ws	= 1 J
1 kgfmeter	kgfm	= 9.81 J
1 footpound	ftlbf	= 1.356 J
1 hphour	hphr	= 2.685E6 J
1 Br.therm.unit	Btu	= 1055 J
1 inchpound	inlbf	= 0.133 J

Thermal Conductivity

1 cal/cms °C		= 419 J/ms K
1 kcal/mhr °C		= 1.163 J/ms K
1 Btu/fthr °F		= 1.73 J/ms K
1 Btu/in/ft^{2}hr °F		= 0.144 J/ms K
1 Btu/fts °F		= 6230 J/ms K
1 W/mK		= 1 J/ms K
1 kilowatthour	kWhr	= 3.60E5 J
1 Calorie	cal	= 4.19 J

Appendix VIII   List of Acronyms

ABDO	Abbreviated Doolittle algorithm
ANOVA	Analysis of variance
CP	Process potential index
CpK	Capability index
CRT	Cathode ray tube
CuSum	Cumulative sum
CV	Coefficient of variation
DAS	Data acquisition system
DC	Direct current
DCU	Data collection unit
DF	Degrees of freedom
DOE	Design of experiments
DOX	Design of experiments
DSC	Differential scanning calorimeter
EVOP	Evolutionary operation
FS	Full scale
GOC	Gravimetric output control
GR&R	Gage repeatability and reproducibility
LCL	Lower control limit
L/D	Length to diameter ratio
LSL	Lower specification limit
MANOVA	Multiple analysis of variance
MFI	Melt flow index
MI	Melt flow index
MPa	Megapascal (10^6 pascals)
MS	Mean squared deviation
PDC	Portable data collector
PID	Proportional/integral/derivative
PMA	Portable machine analyzer
psi	Pounds per square inch
QC	Quality control
R&R	Repeatability and reproducibility
SEC	Specific energy consumption
S/N	Signal to noise ratio SS Sums of squares
SPC	Statistical process control
SSC	Sums of squares of the columns
SSE	Sums of squares of the error

SSR	Sums of squares of the rows
SST	Sums of squares of the total
TC	Thermocouple
UCL	Upper control limit
USL	Upper specification limit

Appendix IX   Proposed New Terminology

D. J. Wheeler has proposed to introduce new terminology for use in statistical process control. His suggestions were published in the newsletter from Statistical Process Controls, Inc. and SPC Press, Inc., No. 2, 1998.

- Instead of *statistical process control,* we should use the term *continual improvement.* This has the advantage of focusing attention on the job of making things better by getting the most out of our current systems and processes as opposed to merely monitoring the process to maintain the status quo.
- Instead of *control chart,* we should use the term *process behavior chart.* This has the advantages of avoiding the baggage associated with the word "control" and correctly describing how to use the chart to get the most out of an existing process.
- Instead of an *in-control process,* we should use the term a *predictable process.* Instead of an *out-of-control process,* we should use the term *unpredictable process.* All too often the words "in-control" are used to describe a situation where all of the product falls within the specification limits. The words "predictable" and "unpredictable" do not carry the same connotations. It is easy to make a distinction between a "predictable process" and "acceptable product."
- Instead of an *out-of-control point,* we should use the term a *point outside the limits.* Instead of an *in-control point,* we should use the term a *point inside the limits.* This simply replaces emotionally loaded terms with descriptive phases.
- Instead of *control limits für individual values,* we should use the term *natural control limits.* Instead of *control limits for averages,* we should use the term *limits for averages* (upper average limit, lower average limit). Instead of *control limits for ranges,* we should talk about *limits for ranges* (upper range limit, lower range limit). These changes are not as hard to get used to as they might seem at first and they avoid the red-herring of "control limits."

Subject Index

α (alpha) risk 190
β (beta) risk 190
$\bar{x}$ and R 131
$\bar{x}$ and s 131. *See* mean and standard deviation
$\bar{x}$ and x 131

A.V. *See* appraiser variation
acceptance sampling 66
accuracy 104
acetal 53
acrylic 53
acrylonitirile butadiene styrene 53
aim of process control 65
alarms for out-of-spec data 99
alkyds 53
alpha risk 190, 195
ambient temperature 73
amorphous plastics 53
analysis of variances 78
animation 216
ANOVA 78. *See* analysis of variance
appraiser variation 112
Archimedean screw 5
artificial intelligence 217
assignable causes 73, 131
asynchronous mode 216
attributes 83
attributes data 87, 131, 140
auto pattern analysis 219
autocorrelation 219
average 67, 76
average and range chart 116
average chart 117

B vs. C analysis 188
backpressure 3
balancing the runner system 18
ball check valve 6
ballpark stage 181
bar chart 80
bar coding 100
barrel 8
barrel heater burnout 73
barrel temperature 73
barrel temperatures 124
barrel wear 73
bayonet-type thermocouple 124
bell shaped curve 84

beta risk 190, 195
Bhote, K. 175
bias 104
bimodal distribution 82
binomial distribution 83
black specks 91
block designs 184
blocked factorial design 184
blush 24
Box, G.E.P. 176
brainstorming session 70, 89

calibration 103
calibration control 106
calibration schedule 106
capable 71
capable process 70
capillary type pressure transducer 122
capping run stage 182
Carter, B. 194
cascade loop 126
cause and effect diagrams 70, 88
cause listing diagram 88
cavity 14
cavity pressure 98
cavity pressure-time curve 98
c-chart 140, 147
central limit theorem 85
central lines 133
central tendency 76
chance causes 73
channel depth 185
chaotic causes 73
check sheets 95
check valve 6
Chris Rauwendaal Dispersive (CRD) mixer 37
clusters 137
coefficient of variation 77
cold runner molds 16
cold slug traps 19
common cause system 101
common causes 73, 131
common causes of variation 176
components search 177, 180
cone-and-plate rheometer 62
consistency index 58
contamination in raw material 73
contiguous solids melting 33

control charts   67, 131
control limits   133
control system   28
controllers   216
conventional SPC   175
conveying   5, 29
cooling time   4
cooling water temperature   98
cooperative action   69
core   14
correlation analysis   94
correlation coefficient   94
correlation table   93
CRD non-return valve   7
creep   54
crosslinking   53
crosslinking reaction   53
crystal polystryrene   54
crystalline regions   53
crystallites   53
crystals   53
CSM *See* contiguous solids melting
cumulative sum (CuSum) chart   132
curing   53
CV *See* coefficient of variation cycle   136
cyclical variation   179

DAS *See* data acquisition system
defective parts   140
defects   140
degassing   5
degradation   29, 62
Deming, W. E.   68
density   62
derivatives   216
design of experiment   175
design of experiments   68, 176, 217
devolatilization   5
die   28
die characteristic curves   30
die forming   29
dimensional variation   89
discrimination ratio   117, 120
dispersed solids melting   33
distribution patterns   78
DOE *See* design of experiment
double toggle clamp   13
dr *See* discrimination ratio
drag flow   30
drag forces   29
DSM *See* dispersed solids melting
dual sensor temperature control systems   126
dynamic process behavior   216

E.V. *See* equipment variation
EDAS *88*. *See* electronic data acquisition system
edge gates   19
effect of the main variables   184
ejection mechanism   14
electrocardiogram of an injection molding process   98
electronic data acquisition system   88
electronic transducers   122
elimination stage   181
engineering limits   70
entanglements   58
enthalpy   61
environment   70
equipment variation   112
evolutionary operation   192
EVOP *See* evolutionary operation
expert systems   217
exponential distributions   83
exponentially weighted moving average   215
exponentially weighted moving average chart   220
extruder barrel   8, 27
extruder characteristic curve   31
extruder screw   6, 27, 29
extrusion   27

factorial analysis stage   181
feed hopper   10, 27
feed opening   10
feed throat casting   28
fill time   98
first-order effects   187
fishbone diagrams   70, 88
Fisher, R. A.   176
fixed station data acquisition systems   95, 98
flash   12, 92
flow rate cooling water   98
flush mounted melt temperature sensors   124
fountain flow   23
fractional factorial   182
freaks   137
frequency curves   78
frequency distribution curve   80
frequency polygon plot   80
frequency polygons   78
frictional heating   32
full factorial experiment   177
fuzzy logic   217

gage linearity   114
gage R&R   109
gage R&R, graphical approach   116
gage R&R, tabular method   115
gage stability   113

Subject Index

gate 2, 19
Gaussian CpK 215
Gaussian distribution 84
goals of SPC 65
GR&R 109. *See* gage repeatability and reproducibility
grand average 133
graphical control chart blocks 216
graphical models 216
graphical representations of a manufacturing Operation 216
green zone 194

heating 5
heating and cooling 28
heating and melting 29
helix angle 185
high degree of shear thinning 58
high density polyethylene 53
high restriction die 31
histograms 70, 78
hold time 2
holding 98
hopper 10
hot runner nozzles 21
hot runner systems 20
human errors 100
hydraulic clamping systems 13

ICC *See* intraclass correlation coefficient
impact resistance 54
impact strength 55
important process parameters 120
individual measurement and moving range chart 131
individual measurements 75
induction time 62
injection forward 98
injection molding cycle 2
injection-molding machine 1
instrumentation 28
interaction 185
interaction effects 184
International System of Units 103
intraclass correlation coefficient 118
Ishikawa diagrams 88

jetting 24

kurtosis 82

Latin square 184
liquid crystalline plastics (LCP) 54
log normal distribution 83

logarithm 84
logic networks 204
long term precision 109
lopsided distribution 83
low density polyethylene 53
low restriction die 31
lower control limit 116
lower control limit for the c-chart 144
lower control limit for the np-chart 144
lower control limit for the p-chart 143
lower control limit for the range chart 133
lower control limit for the u-chart 147
lower control limit for the x-bar chart 133
lower specification limit (LSL) 66

machine 70
Maddock, B. 32
man 70
material 70
maximum injection pressure 2
mean 67, 75
mean, $\bar{x}$ and sigma, s, chart 132
mean, $\bar{x}$ and range, R, chart 132
measurement 70, 103
measurement error 70, 106, 117
measurement system 70
measurement system evaluation 87, 114
measurement variation 103
mechanical clamping systems 12
median 76
median, $\bar{x}$ and individual measurement, x, chart 132
median, $\bar{x}$ and range, R, chart 132
melamines 53
melt index 56
melt index tester 56
melt temperature 124
melting 33
melting point 54
mercury 122
methods 70
metrology 103
MFI 56. *See* melt flow index
MI 56
milieu 70
mixing 5, 28, 101
mixture 138
modality 82
mode 76
mold 12
mold filling analysis 26
mold open 97
mold temperature versus warpage 93
molecular weight 56

monitoring systems 98
Monte Carlo simulation 177
morphology 54
multimodal distributions 82
multiple environment overstress test 175
multi-van (MV) chart 179

National Institute of Standards and
    Technology 103
natural causes 73
negative correlation 94
negative skew 83
nested designs 184
neural network technology 217
neural networks 59
Newtonian fluid 75
nominal 83
nonconforming products 71
nonconforming units 140
nonconformities 140
non-Gaussian CpK 215
non-linear fluid 58
non-linearity 105
non-Newtonian fluid 58
nonparametric comparative experimentation 188
non-rational sampling 101
non-return valve 6
normal distribution 78, 83
nozzle 8
np-chart 144
null hypothesis 188
nylon 53

object-oriented modeling 203
observed variation 103
OEMs See original equipment suppliers
one-at-a-time approach 185
one-at-a-time experimentation 186
opaque 54
operational thumbprint 98
operator error 106
optical pressure transducer 123
original equipment suppliers 98
orthogonal arrays 177
orthogonal statistics 187
out of calibration 104
overtoggling 14

P/T See pressure/temperature
paired comparison 177
paired comparisons 182
Pale Pink X 179
Pareto charts 70
Pareto diagrams 91

Pareto plot 216
part variation 119
Pascal·second 55
pattern observation blocks 217
P-C lines See precontrol lines
p-chart 141, 142
PDCs See portable data collector
phenolics 53
piezoelectric transducers 124
piezoresistive pressure transducers 123
Pink X 179
pinpoint gate 12, 19
plain tip 7
plant-wide monitoring systems 99
plastic melt 53
plastics 53
platen 1
PMAs See portable machine analyzers
Poise 55
Poisson distribution 83
polycarbonate 53
polyester terephthalate 53
polymers 53
polypropylene 53
polystyrene 53
polyvinylchloride 53
population 75
portable data collectors 95, 97
portable machine analyzers 97
positional variation 179
positive correlation 93
positive skew 83
power law equation 58
power law fluid 58
power law index 58
precision 104
precontrol 194
precontrol lines 194
pressure relief device 122
pressure, effect on viscosity 59
pressure/temperature (P/T) transducers 123
pressure-time profile 122
primary reference standards 103
process analysis diagram 88
process capability 70, 131
process control charts 70
process drift 217
process parameters 120
process problems 217
product control 66
product specifications 70
product tolerance 106
product variability 71
product variation 117

Subject Index

production summaries 99
profits 69
projected area 12
Pro-T-coN 216
pseudoplastic behavior 58
Purcell, W. 194
pushrod type pressure transducer 122
PVT diagram 64

quality assurance (QA) 217
quality control 67
quality function deployment 175

ram extruder 56
ram-screw 5
random causes 73
random number generators 184
random number tables 100
randomization 190
range 67, 76
range chart 116
rate of shearing 57
rational subgroups 75
realistic tolerance parallelogram plots 192
realistic tolerances 193
real-time monitoring 87
real-time SPC 217
recipes 99
reciprocating screw extruder 5
recipro-screw 5
recovery 98
recycling 53
Red X 179
red zones 194
regression analysis 217, 218
relative humidity 73
repeatability 107, 114, 118
repetitive pattern 137
replication error 117
reproducibility 107, 116
residual error 184
resolution 105, 127
resultants 216
ring check valve 6
rod-like structures 54
root-cause analysis 217
RS extruder 5
run 137
runner system 16
running record 117
rupture disk 122

sample 75
sample size 101

sampling 101
Satterthwaite, F. 194
scatter diagrams 70, 93
scatter pbots 192
screw position 97
screw speed 126
screw speed measurement 126
screw tip 7, 122
screw velocity 98
screw wear 73
secondary reference standards 103
self-directed work teams 175
semi-crystalline plastics 53
sensitivity 105, 126
separating force 12
setpoint generator 217
setting tolerances 192
Shainin methodology 177
Shainin, D. 68, 176, 177, 194
shear rate 32, 57
shear thinning behavior 58
Shewhart, W.A. 68, 175
Shewhart control chart 67
shift in level 138
short term precision 109
shrinkage 55
shut-off nozzles 9
SI system 103
signal to noise ratio 77
single toggle clamp 12
sink marks 91
skew 83
SMI *See* standard manufacturing instructions
soak time 3
SPC *See* statistical process control
spec *See* specification
special causes 73
special causes of variation 176
specific energy 61
specific heat 61
specific volume 63
Specification limits 65
specification width 71
spiral length tester 57
splay 92
sporadic causes 73
spread of the data 78
sprue 8, 15
sprue gates 19
stability 105
standard deviation 76
standard deviation of the measurement error 108
standard deviation of the repeatability 107
standard deviation of the reproducibility 108

standard manufacturing instructions 216
statistical methods 65
statistical process control 65
statistics 65
sticking nozzle valve 73
stratification 101, 138
Streamer1 215
strong negative correlation 94
strong positive correlation 93
subgroup 75
subgroup size 133, 141
symmetrical distribution 83

Taguchi 176
Taguchi loss 215
target dimension 75
TC *See* thermocouple
temperature coefficient 59
temperature control 59
temperature, effect on viscosity 59
temporal variation 179
tertiary reference standards 103
test-retest error 117
TGA *See* thermogravimetric analyzer
thermal conductivity 61
thermal insulators 61
thermal properties 60
thermal stability 62
thermocouple 125
thermogravimetric analyzer 62
thermomechanical analyzer 62
thermoplastic elastomers 55
thermoplastics 53
thermoset rubber 55
thermosets 53
three Sigma control limits 140
three-factor interactions 187
tie bars 1
time-series predictions 215
TMA 62
toggle clamp system 12
tolerance 106, 191
tooling 54
tooling adjustment 73
torque 56
TPE *See* thermoplastic elastomer
traceability 100, 103
traditional gage R&R 115
training 70
transcription errors 100
translucent 54
trend 137
trending 99

true value 104
true variation 103
tunnel gates 19
two-factor interactions 187
two-plate mold 16
two-stage extruder screw 5

u-chart 140, 145
unimodal distribution 82
units of measure 103
universe 75
unnatural causes 73
upper control limit 116
upper control limit for the c-chart 146
upper control limit for the np-chart 144
upper control limit for the p-chart 141
upper control limit for the range chart 133
upper control limit for the u-chart 147
upper control limit for the x-bar chart 133
upper specification limit (USL) 67
ureas 53

variability 66
variables 83
variables data 87, 131
variables search 178, 182
variance 76, 182
variation analysis diagram 88, 89
variation within each unit 180
vent opening 5
vented extruder 5
venting 24
viscosity 55
viscous heat generation 32, 60
viscous heating 32
voids 92
voids in a product 89
volatiles 5

warpage 54
weak negative correlation 94
weak positive correlation 93
weld line 25
working standards 103

x and Rm 131
x-bar and R chart 67

Yates, F. A. 176
Yates algorithm 188
yellow zones 194

zone rules 137